Introduction to Crystallography

Frank Hoffmann

Introduction to Crystallography

 Springer

Frank Hoffmann
Universität Hamburg
Hamburg, Germany

The book was originally published in German under the title "Faszination Kristalle und Symmetrie", 1. Auflage, Springer Spektrum, © Springer Fachmedien Wiesbaden GmbH 2016.

ISBN 978-3-030-35112-0   ISBN 978-3-030-35110-6   (eBook)
https://doi.org/10.1007/978-3-030-35110-6

This Springer imprint is published by the registered company Springer Nature Switzerland AG
The registered company address is: Gewerbestrasse 11, 6330 Cham, Switzerland

Preface to the First English Edition

Dear Readers,

This textbook on crystallography is a translation of my German textbook *Faszination Kristalle und Symmetrie* that was published in 2016. I want to thank Svetlana Zakharchenko and Charlotte Hollingworth of Springer for their continuous support and for realizing this project!

Apart from correction of some minor formal errors and updating some references, the text is unchanged.

Now, the English version of the textbook on crystallography is in front of you. You are invited to take a round trip through forms, patterns, symmetries, and structures. Excursuses about the reception of beauty will play just as an important role as specific structural considerations.

It is not like there is a gap in the textbook market for crystallography. The decision, to add yet another one to the existing textbooks, was based on a teaching experience from 2014 to 2015: I offered a 7-week MOOC ("Massive Open Online Course") on the iversity platform on the subject of crystals and symmetry. Approximately 2,000 participants were enrolled. As in a normal lecture series at a university, the interest declined with increasing degree of difficulty, but the decisive factor was that there were participants from all over the world, with very different professional, cultural, and socioeconomic backgrounds. And everyone shared the same fascination for crystals! Also, in the accompanying forum to the course, a stimulating exchange developed with a total of almost 500 contributions. This course was aimed at beginners who are interested in the area but had otherwise no prior knowledge on the subject. The present book follows the same approach. The content is presented in a way that you should be able to comprehend the matter with a minimum level of mathematics and a minimum number of formulas.

For me as a lecturer, the most important thing is to try to make the *basic concepts* of a subject clear. This should be done in such a way that the quasi-natural curiosity of the learners is preserved as long as possible. The quantitative and mathematical relationships have been written down countless times; they certainly do not require a new book. And to be honest, I have doubts that the fact that you can express the symmetry operation of a horizontal reflection also as a matrix would add anything to the *understanding* of that operation:

$$\begin{pmatrix} 1 & 0 & 0 \\ 0 & 1 & 0 \\ 0 & 0 & -1 \end{pmatrix}.$$

In these times, with all the possibilities of modern and interactive e-learning, YouTube channels, etc., a *book* seems a bit old-fashioned. From this perspective, it might be quite unsuitable for the intended purpose. So, this book can only be the imperfect result of an attempt to involve the readers as much as possible. It is a book that is intended to encourage active cooperation not only while reading it but also beyond: you are invited to make crystal models from paper and to discover their symmetry properties, and, if applicable, you can photograph your own crystals or minerals and learn how to classify them according to their crystallographic point groups. Furthermore, there are some crystallographic brain teasers included. Additionally, you can explore the three-dimensional atomic structure of many minerals using freeware on the computer screen. You can find all accompanying materials at the following URL:

▶ https://crystalsymmetry.wordpress.com/textbook/

Here, you will also find the possibility to contact me, if you have any further questions.

I wish you a lot of fun while going through this book, and I would be pleased if your interest and fascination about crystallography will have even grown afterwards!

Frank Hoffmann
Hamburg, Germany

Contents

Introduction

© Springer Nature Switzerland AG 2020
F. Hoffmann, *Introduction to Crystallography*, https://doi.org/10.1007/978-3-030-35110-6_1

1

1.1 Crystals, Structures, Orders of Magnitude

In this very first chapter, I would like to give you an idea of the term "crystal structure." First of all, we want to get an impression of how large or small crystalline objects can be. For this purpose, take a look at ◘ Fig. 1.1, which is based on the popular short film "Powers of Ten" from 1977 [1]. This short film illustrates in a beautiful way, of which sizes objects are that are subject of scientific investigations. They span an impressive range of orders of magnitude. We do not want to explain all objects in detail, but instead will focus on just a few.

Let us start with the largest objects, astronomic objects. Galactic objects do have a structure as the universe itself has a particular structure and, interestingly, a very heterogeneous one. On the one hand, there are space areas, which are very dense, for instance, globular clusters or super globular clusters, which consists of typically 100,000 stars located in a relatively small area. On the other hand, there are huge areas, which are almost empty. The topic of the universe belongs rather to the discipline astrophysics and is only in one particular aspect related to crystallography: certain minerals were found in meteorites which do not occur on earth. The examination of these meteoritic minerals is a very interesting area within mineralogy and is able to deliver useful insights into the early history of our universe [2].

When going down a few orders of magnitude in ◘ Fig. 1.1, we will first pass planet earth and then will reach a house with an approximately length/height of 10 meters. What do you think: do crystals exist, which are as large as houses? This is indeed the case – look at the spectacular photograph in ◘ Fig. 1.2!

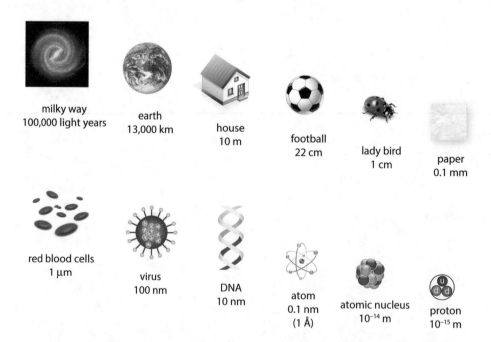

milky way
100,000 light years

earth
13,000 km

house
10 m

football
22 cm

lady bird
1 cm

paper
0.1 mm

red blood cells
1 μm

virus
100 nm

DNA
10 nm

atom
0.1 nm
(1 Å)

atomic nucleus
10^{-14} m

proton
10^{-15} m

◘ **Fig. 1.1** "Powers of Ten": Objects of different size orders of magnitude (after a short film by Charles and Ray Eames from 1977 [1]. (House: CC BY 3.0 ► webdesignhot.com, football, red blood cells, virus and DNA: CC BY 3.0 designed by ► freepik.com, lady bird: CC BY 3.0 ► freedesignfile.com, sheet of paper: CC BY 3.0 ► vector4free.com)

Fig. 1.2 Gypsum crystals in a cave of the mine of Naica (Mexico). Note the researcher in the bottom right for size relations. (CC BY 3.0 Alexander Van Driessche)

You can see a researcher examining huge crystal pillars consisting of selenite, a variety of the mineral gypsum, calcium sulfate, $CaSO_4 \cdot 2\,H_2O$. This crystal cave (which is named "*Cueva de los cristales*") is located in Mexico and is a side cave of the mine of Naica, which was discovered by accident only in the year 2000. The exploration of this cave is not an easy task due to the drastic conditions with temperatures up to 50 °C and a relative humidity of almost 100%. For this reason, the researchers also have to wear special cooling suits; still they are not allowed to work for more than ~ 1 hour; otherwise a life-threatening circulatory breakdown may occur. The exploration of this cave is an exciting story, which has been told in various formats and channels [3, 4]. A scientific treatise concerning the crystal nucleation and growth of these gigantic crystals can be found in refs. [5, 6].

Obviously, the physicochemical conditions in that cave were constant over a very long period of time allowing the crystals to grow to their actual size, the largest reaching 14 meters. According to estimates, the largest crystals are 100,000 to 1 million years old.

Let's switch back to the overview in ■ Fig. 1.1. If we go down further, we are reaching objects of the size between a football and a ladybird. This is the typical size of mineralogical objects which are on display in exhibitions or museums. In ■ Fig. 1.3 a deep green fluorite crystal (also called fluorspar, calcium fluoride, CaF_2) is shown. Usually fluorite is colorless and transparent; however, this item is colored due to impurities. This fluorite crystal has a size of approx. 2 cm and is set on a matrix of mica (a particular family of layered silicates).

Going down the scale of orders of magnitude another step we are reaching the thickness of a sheet of paper, i.e., approx. 0.1 millimeter. This is the typical size of single-crystals obtained in chemical research labs. In ■ Fig. 1.4 a lovely blue sparkling crystal of a so-called metal-organic framework is shown, a chemical compound class which we will get

1

◻ **Fig. 1.3**   Green fluorite crystal with a size of approx. 2 cm from Namibia. (CC BY-SA 3.0 CarlesMillan)

◻ **Fig. 1.4**   Single-crystal of a copper-based metal-organic framework; edge length approx. 0.03 mm

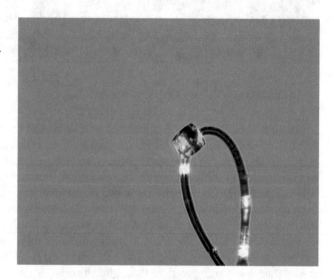

acquainted with in ▶ Chap. 8. This crystal is fixed with a nylon loop in preparation of an X-ray diffraction experiment, in order to resolve its atomic structure.

Once again back to the overview in ◻ Fig. 1.1: to describe the outer shapes of crystals, their morphology (see ▶ Chap. 2), we deal with sizes in the range of millimeters to centimeters. But crystals do, of course, possess also an inner structure. And in order to describe this inner structure, we have to go down another few steps along the scale of orders of magnitude until we reach the tiny dimensions of molecules and atoms. Atoms or molecules are the building blocks of crystals, which are arranged in a specific manner that we will explore shortly.

We could go down a few more steps and would reach subatomic structures and dimensions. Atoms as well as their atomic nuclei, consisting of protons and neutrons, do have a structure. But clearly, the exploration of these structures belongs to the domain of high-energy physics which we will not deal with further in this book.

The Term Structure from a Linguistic Point of View

The term structure is derived from the Latin noun *structura*, which means "fitting together, adjustment, building," and the verb *struere*, meaning "to pile, build, assemble." This term was used until the seventeenth century almost exclusively in the architectural area in the sense of "construction," for instance, of a house. However, with a progressive acquisition in knowledge in natural science, linguistics, and medicine, the word was later also and is now still used in many different fields in the sense of state, texture, quality, property, condition, etc. This can be, for instance, the structure of tissue (biol./med.), the structure of language (grammar), the constitution of substances (chemistry), the structure of crystals (mineralogy), or the fabric of materials.

Summary: If we want to describe the inner structure of crystals, we are in size range of 10^{-10} meters (in chemistry this length unit is known as 1 Ångström, named after the Swedish physicist Anders Jonas Ångström, 1 Å = 0.1 nanometer) and a few nanometers; and if we want to characterize the outer shape of crystals, then we are in a size region between a sheet of paper and a football, the giant crystals of the mine of Naica being an exception.

1.2 Specialist Terms: Crystallographic Poetry

Every scientific discipline uses special terms, foreign words, particular terms which may sound at first bizarre and which we simply have to memorize. However, according to the author's own impression, the vocabulary in crystallography is particularly strange. You would be forgiven for thinking that crystallographers speak a cryptic language. Therefore, the author decided to take some of these special terms to put them together in something, which might be more appealing – a little poem. It is only for entertainment purposes, and don't be afraid, all terms will be explained step-by-step afterward. A spoken variant accompanied with music can be found at the book's website at the internet:
► https://crystalsymmetry.wordpress.com/textbook/

» 1bar | just another name for inversion centre | 2,3,4-fold | 6bar
base | but where is the superstructure? | that is the motif
primitive | face or body-centered | give rise to various lattice types | explored by Bravais
there are holes in a coordination polyhedron | tetrahedral and octahedral
cube, prism, rhombic dodecahedron | corner or edge connected?
the cell is fundamental | it can be skew, triply or once | triclinic or monoclinic
if the symmetry rises | it can be orthorhombic, rhombohedral, tri- or hexagonal
or even tetragonal or cubic | these are the crystal systems | there are 7 of these
the dress of crystals is made by their faces |
their relative sizes are called habitus | altogether: the morphology |
classified according to point symmetries | are nothing more than a crystal class
Finally, the symmetry is described completely by the central space group
same mineral, different minor constituents, different habitus
not a variant but a variety, often isotypical
Mr. Miller expresses indices as *hkl* values
which denominate the lattice plane families, for instance 111
And if, finally, a crystal is not a regular crystal, then it is not
quasi crystalline but only a quasicrystal…

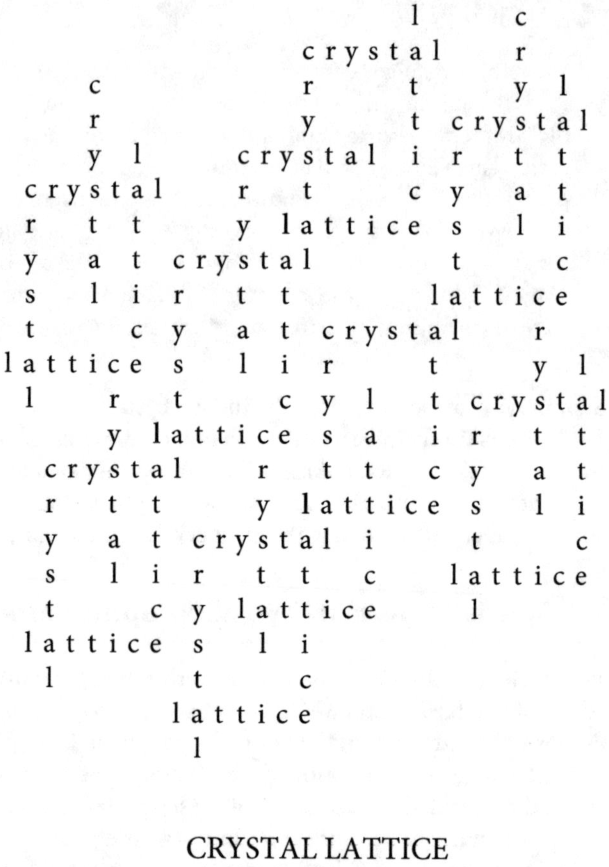

1

Fig. 1.5 Experimental poetry with a crystallographic background: Out of the two words crystal and lattice, a crystal lattice of characters is built. (Crystal Lattice from „Crystallography", 2nd revised edition, © Christian Bök, 2003. Courtesy of Coach House Books)

```
                                              l        c
                                 crystal         r
                    c              r      t         y  l
                    r              y         t  crystal
                    y  l        crystal  i     r      t     t
          crystal         r      t         c     y     a     t
          r     t     t         y  lattice  s        l  i
          y     a  t  crystal                t           c
          s     l  i     r     t     t              lattice
          t        c  y     a  t  crystal              r
     lattice  s     l  i     r           t        y  l
          l        r     t        c  y  l     t  crystal
               y  lattice  s  a        i     r     t     t
          crystal         r     t  t     c  y        a  t
          r     t     t         y  lattice  s        l  i
          y     a  t  crystal  i           t           c
          s     l  i     r     t     t     c        lattice
          t        c  y  lattice              l
     lattice  s     l  i
          l              t        c
               lattice
               l
```

CRYSTAL LATTICE

So far, a non-conventional overview of some of the topics covered in this book. In a far more elegant and linguistically erudite way, the Canadian, internationally renowned experimental poet Christian Bök delves into the area of crystallography. In his book *Crystallography* [7], he suggests to consider language as such as a crystallization process – in fact, a very interesting idea. As an example, in one of his experimental works, he arranged the two words "crystal" and "lattice" in such a way that they both build a crystal lattice of letters (see Fig. 1.5).

But at this point our small journey into the art of poetry should come to an end, and in the next section, we will approach the term crystal from a crystallographic point of view.

1.3 The Term Crystal and Its Definition

One way to approach a subject is to look at the word and its original meaning. The word crystal derives from the Greek word "*krystallos*", meaning "ice." Well, ice in its form of snowflakes or glaciers are indeed crystalline; however, not all crystalline materials consist of ice. The term ice was used in a figurative sense and goes back to the very first findings

■ **Fig. 1.6** Quartz crystals (Tibet). (CC BY-SA 2.5 JJ Harrison)

of rock crystals. In nature, rock crystals can relatively easily be found and if you look at ■ Fig. 1.6, you will see that they also look somewhat like ice. In ancient times it was also thought that they were developed in freezing cold. Today we know that they are formed in large heat and under pressure.

Rock crystals are a colorless variety (see background information) of quartz, a mineral with the chemical composition SiO_2, silicon dioxide. There are a great number of varieties of quartz in nature, and many of them show interesting colorings. Let us look at three examples.

Variety

In mineralogy differently developed minerals with respect to its color, transparency, habitus, or size are known as varieties. They often have distinct names. In contrast to modifications, which constitute minerals on its own, the principal crystal structure of varieties is identical. However, the chemical composition of varieties may vary in small amounts. For instance, the chemical composition of the mineral corundum is Al_2O_3. In its pure form, it is colorless. However, a small addition of chromium leads to the well-known variety ruby, which is red.

The *amethyst* is a violet variety of the mineral quartz (see ■ Fig. 1.7, left). The coloring is due to iron ions in combination with irradiation of X-rays or gamma rays. However, the exact structural details which cause the coloring are still a matter of debate. Interestingly, the color of amethysts will fade relatively fast under the irradiation with sunlight.

◘ Fig. 1.7 Three varieties of quartz: amethyst (left) (CC BY-SA JJ Johnson), milky quartz (middle), and rose quartz (right)

In colorless or white *milky quartz*, no metal ions are incorporated, but tiny inclusions of liquid droplets and/or fissures which cause the light to scatter lead to its typical milky-dull appearance.

The rose color is eponymous of the *rose quartz*, which comes in two variants, a dull and a clear one. According to more recent research, the color of the dull rose quartz is due to inclusions of the foreign mineral dumortierite, an aluminum boron silicate, which itself is colored through the incorporation of iron or titanium. The color of the clear rose quartz on the other hand is caused by the coupled incorporation of aluminum (group 13) and phosphorous (group 15) in combination with irradiation of ionizing radiation (often by the presence of the radioactive isotope ^{40}K).

Independent of what the ancient Greeks thought about crystals, there is a clear definition:

> Crystals are homogeneous, anisotropic solid states, whose building blocks are strictly three-dimensional periodically ordered.

This last attribute – periodically ordered – is the most important one. It is the absolute feature of the crystalline state of matter. There are other solid-state materials, which do not have such perfect order of its constituents, such as wood, glass, wool, plastic, etc.; they are amorphous. "Amorphous" means literally "without shape," which might be confusing at first glance, because a particular wood block or plastic bottle does possess a certain shape, which it also keeps. The attribute amorphous, therefore, refers to the inner structure of these materials, which are not regularly ordered. In general, all compounds can be categorized according to their aggregate state of matter into (i) solid states, (ii) liquids, and (iii) gases. And the solid states can in turn be again subdivided into the main categories (i) amorphous and (ii) crystalline.

In some textbooks an alternative definition of crystals is given: crystals are solid-state compounds, which possess a crystal structure (see ◘ Fig. 1.8). This might not be particularly helpful, if the meaning of the term crystal structure is unknown. However, together with the definition given above, a first intuitive comprehension is probably attained. The term will be further explained during the course of this book.

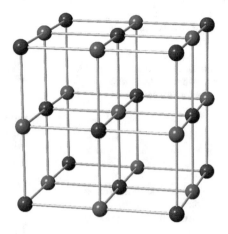

◻ **Fig. 1.8**   Crystals do have a crystal structure

1.4 Anisotropy

All crystals show a feature called anisotropy. This means that certain properties in their magnitude are dependent on their direction; they are *directional*. The opposite would be isotropic and means, conversely, that these properties are the same in all directions or all orientations, for instance, with reference to a particular operation. Examples for anisotropic properties are hardness, cleavability, elasticity, or thermal expansion properties. If we take hardness as an example, it could mean that a crystal block on which you exert a certain pressure is indented to a different extent, depending whether you exert the pressure from above or from the side, respectively (see ◻ Fig. 1.9).

Other anisotropic features can be the magnetization, the electric polarizability, or the electric or thermal conductivity. For instance, in graphite the electric conductivity is 10,000 times larger along the single graphene sheets (in this direction the specific conductivity amounts to $2.6 \times 10^4 \, \Omega^{-1} \, cm^{-1}$, reaching almost metallic characteristics) compared to the direction perpendicular to these sheets. Equally, graphite also shows a far larger tendency of cleavage parallel to the graphene sheets than perpendicular to that direction. We will have a closer look at the structure of graphite in ▶ Sect. 7.4.

However, anisotropy is not an exclusive feature of crystals. There are materials with anisotropic characteristics, which are not crystalline. A familiar example is paper, which can be teared apart much more easily along the cellulose fibers compared to the direction perpendicular thereto.

The exact reason or understanding for anisotropic behaviors might be difficult, in particular regarding the exact quantitative ratios, but the origin of an anisotropic property and the principle behind it can be relatively easily illustrated in a simple picture. ◻ Figure 1.10 shows a two-dimensional, periodic arrangement of spheres. Now we will look along two different directions and the respective sequence of spheres. In direction 1 we see an alternating arrangement of two red spheres and a blue sphere, all lying adjoined

1

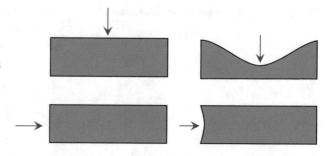

Fig. 1.9 Hardness is an example of an anisotropic property: the same pressure being applied onto the top surface and on a side face of this squared stone leads to very different indentations

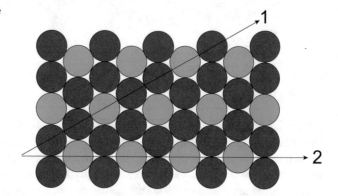

Fig. 1.10 Illustration of the reason for anisotropy: displacement of spheres along direction 1 or 2 requires different forces due to the different relative position and sequence of spheres

in the same line. Along direction 2 the sequence is again two red spheres, one blue sphere, another two spheres, and so forth, but this time the pairs of red spheres are oriented perpendicular to that direction. The blue spheres lie in the indentation built by pairs of the red spheres. If we now imagine trying to push the spheres along direction 1 or 2 against each other, it should be evident that we have to apply different forces.

1.5 Robert Hooke and the Correspondence Principle I

At this point, we already know the definition of crystals, we have a certain understanding of the meaning of the term crystal structure, we have discussed the dimensions of crystals, and we have already looked at a few crystal shapes. In this section we want to bring some of these aspects together with the help of the so-called correspondence principle.

The correspondence principle states that there is a certain relationship between the outer shape of a crystal on the one hand and its inner structure on the other. The question is: which kind of relationship? Is there a law which would allow deducing the inner structure and its constituents of a crystal out of its outer shape?

In ☐ Fig. 1.11 three different schematically drawn crystal shapes are shown. Are we able to deduce from these outer shapes what is inside, to view behind their "curtains"?

Fig. 1.11 Three different crystal shapes. Is it possible to deduce their inner structure out of the outer forms?

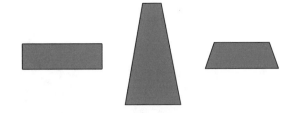

Fig. 1.12 According to the concept of Robert Hooke, the outer crystal forms are the result of differently arranged spheres

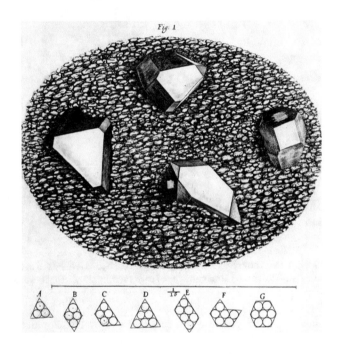

This question is in fact a very old one, and in this respect the author wants to bring the extremely interesting monograph "Micrographia" by Robert Hooke from the year 1664 to your attention [8]; Robert Hooke is known particularly because of his famous Hooke's law ("…the extension of a spring is proportional to the force…"). In this historic era, the light microscope was invented, and many scholars began to investigate tiny objects that formerly could not have been seen by the naked eye, for instance, tissue cells, arthropod eyes or antenna of insects, and many more. Robert Hooke examined crystals, drew impressive sketches of them, and hypothesized: crystals are composed of smaller and identical building units, namely, spheres, and specific arrangement of these spheres lead to different outer crystal shapes (see **Fig. 1.12**).

For that time, taken into account the very limited knowledge concerning the constitution of matter, it was a brilliant idea. His hypothesis was not completely right, but almost: the only misassumption was that the smaller building blocks are spheres. To anticipate the right solution, these building blocks are confined by flat planes, which are enclosed by

1

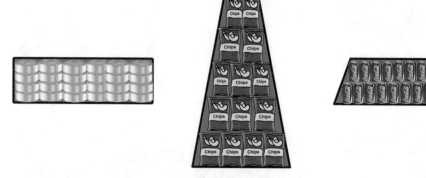

⬛ **Fig. 1.13** The "crystals" of ⬛ Fig. 1.11 consist of tuna tins, potato chips bags, and coke cans

(a)   (b)   (c)

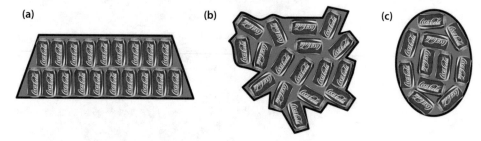

⬛ **Fig. 1.14** The crystal-like arrangement of coke cans is reflected in the sharp boundary of the outer form (**a**). In (**b**) and (**c**) arrangements of coke cans are shown that have no resemblance to crystals (see text)

skew or right angles: the unit cells (see ▶ Sect. 1.7). For all of you, who want to learn more about the early time of the scientific discipline crystallography, the author recommends the excellent review paper "Chemical Crystallography before X-ray Diffraction" from the journal *Angewandte Chemie International Edition* [9].

Let's switch back to the schematically drawn crystal shapes of ⬛ Fig. 1.11. In ⬛ Fig. 1.13, we see of which principle building blocks they were formed. The first consists of tuna tins, the second of chip bags, and the third one of coke cans. Of course, this was not foreseeable, and you might think that the correspondence principle does not hold at all. The problem is that there is no one-to-one relationship between the shape of the single building block and the outer shape of the whole crystal. This is the reason why it is formulated only as a loose correspondence *principle*. What that could mean is illustrated in ⬛ Fig. 1.14a–c. In (a) a regular array of coke cans is shown, which resembles a crystal and its blue enveloping boundary does indeed correspond with this arrangement. On the contrary, it should be immediately clear that the chaotic pile of coke cans shown in (b) does not represent a crystal-like structure and that, therefore, a coke-can-crystal can never have this blue outer shape. Likewise, a crystal can also never possess a globular, spherical, or balloon-like outer shape as shown in ⬛ Fig. 1.14c. This is not possible because there is no regular, periodic array that would lead to such shapes. The last statement might be confusing for some of

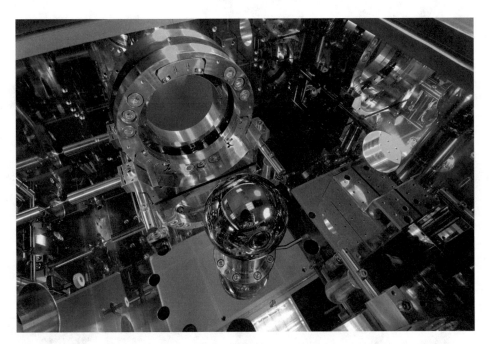

◘ Fig. 1.15 The sphere interferometer of the PTB (Physikalisch-Technische Bundesanstalt), which is able to measure the diameter of the elaborately manufactured single-crystal sphere of silicon with a precession of a few nanometers (Courtesy of PTB)

the readers, because they think of rounded semiprecious or gemstones or worry stones. Others of you might have been heard about the so-called Avogadro project, which aims at defining the mass unit kilogram according to a new basis. For this purpose, an almost perfect sphere consisting of a silicon single-crystal is needed (◘ Fig. 1.15). This should substitute the international prototype kilogram that is stored in a safe at the International Bureau of Weights and Measures in Sèvres near Paris. Simultaneously, this silicon sphere should serve as a new basis for the definition of the unit of the amount of substance (the mole) [10]. The explanation for these round crystals is simple: these are not their natural shapes to which they grew. Instead these are round-sanded forms, either by natural processes in the case of some rock-building minerals or by (highly precise working) technical apparatuses in the case of the silicon sphere.

Let us look at an example of a real crystal in its natural form. In ◘ Fig. 1.16 a non-sanded item of a quartz crystal in its natural form is shown. To be precise this specimen belongs to the violet variety of quartz; it is an amethyst.

Quartz consists of pure silicon dioxide, and the building blocks are SiO_4 tetrahedra, shown in ◘ Fig. 1.17a, which are exclusively corner-connected to build the whole crystal. However, the crystal of ◘ Fig. 1.16 does not look like a tetrahedron. And also a larger piece of the atomic structure, this of the unit cell (see ▶ Sect. 1.7) as shown in ◘ Fig. 1.17b, does not resemble the outer shape. However, if we consider the whole crystal, we see that the atomistic ensemble consisting of silicon and oxygen atoms shown in ◘ Figure 1.17c reflects exactly the shape of the real quartz crystal in ◘ Fig. 1.16.

Fig. 1.16 Single-crystal of amethyst on a matrix of calcite from Afghanistan, size approx. 10 mm. (Courtesy of A.C. Akhavan, © 2005–2013)

(a)   **(b)**   **(c)**

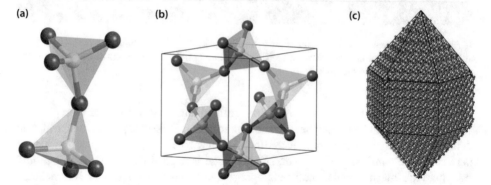

Fig. 1.17 **(a)** The building blocks of quartz are SiO_4 tetrahedra, which are exclusively corner-connected; here two such tetrahedra are shown. **(b)** Unit cell of quartz. **(c)** Atomistic model of the single-crystal shown in ◨ Fig. 1.16 (Note that the size relations are not realistic; this model would be only 11 nanometers in size and not visible, even under a microscope)

We see that there is indeed a relationship between the inner structure and the outer shape of a crystal; however, the question is always on which scale this relationship will manifest itself. We will come back to that aspect in a later stage in this book. For now, we want to consider a further example, in which the correspondence principle is more evident. In the next section, it is getting frosty.

1.6 The Hexagonal Symmetry of Snowflakes

Water in its liquid or solid state, for example, in the form of ice, snow, corn snow, etc., probably belongs to one of the most fascinating forms of matter at all. Water exhibits dozens of so-called anomalies; this means that certain macroscopic properties are different to what is expected or show different behaviors compared to other substances. The best-known example

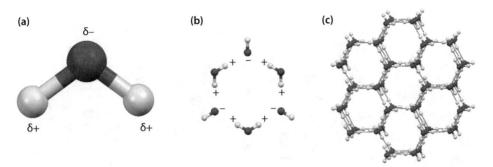

Fig. 1.18 (a) Atomistic model of water with positive and negative partial charges at the hydrogen and oxygen atoms, respectively. (b) Ensemble of water molecules which are oriented in such a way that positive partial charges are directed to negative ones and vice versa; H-bonds are formed. The hydrogen atoms are partially covered; they are also connected in the third dimension perpendicular to the drawing plane. (c) The prevailing hexagonal shape of snow crystals appears

is the fact that water *expands* below 4 degrees Celsius when the temperature is sinking further, whereas almost all other substances will shrink continuously with falling temperature.

Water, H_2O, is a three-atomic, angled molecule (see ◘ Fig. 1.18a); the red sphere represents an oxygen atom, while the white spheres represent hydrogen atoms. Water is a highly polar molecule: the oxygen atom carries a negative partial charge and the hydrogens atoms a positive one. If several water molecules aggregate, they orient in such a way that always a positive charge is directed to a negative and vice versa – charges with opposite signs are being attracted to each other (see ◘ Fig. 1.18b) – so-called hydrogen bonds are formed. If more and more water molecules aggregate, the growing solid cluster will develop a discernibly visible outer hexagonal shape. In ◘ Fig. 1.18c a respective mini ice crystal is shown as an atomistic model. This hexagonal shape is not a product of coincidence – all snowflakes or ice crystals of this planet are hexagonal!

In ◘ Fig. 1.19 some wonderful photographs of single snow crystals are gathered. By the way, a typical snow crystal consists of approx. 10^{18} (one quintillion) water molecules. The individual snow crystals might look different in detail; however, they share one thing in common, namely, their hexagonal symmetry; they possess a six-fold axis of rotation (a rotation of 60° does not change their appearances; for a basic introduction into the topic symmetry, see ► Chap. 3). At the fantastic website of Prof. Kenneth G. Libbrecht (► snowcrystals.com), which is explicitly concerned with the science of ice and snow, you can find further beautiful photographs of snow crystals. On this page you can also get an answer to the famous question, if really no two snowflakes are alike, including a footnote, if this is a meaningful question at all.

Some of the snow crystals that can be found at the ► snowcrystals.com website look like exemptions, and they do not look like hexagons; rather they have a very thin, needle-like appearance (see ◘ Fig. 1.20). However, also these snow crystals have a hexagonal symmetry: if we look from above or from the bottom along the long axis of this needle, we will see that these are hollow needles, whose inner walls do possess a hexagonal cross section.

Why do the snow crystals in ◘ Fig. 1.19 look so different apart from their common six-fold rotational symmetry? Because they grew under different conditions. Their specific forms – in a recent publication no less than 121 principally different shapes were identified [11] – are mainly dependent on the temperature and relative humidity of the air in which the snow crystals grow (see ◘ Fig. 1.21). Interestingly,

1

⬛ Fig. 1.19   A selection of very differently formed snow crystals. (Courtesy of Kenneth G. Libbrecht)

the details concerning the growth process are complicated to such an extent that they are not fully understood, yet. Articles concerning the physics of snow crystals that are worth reading can be found in refs. [12–14]. This is one of the reasons why research on snow, ice crystals, and the different ice modifications is still a fast-developing area. In this respect the question concerning the smallest possible snowflake is worth mentioning. How many water molecules have to aggregate to something that can be reasonably considered as a snow crystal? The answer was given not long ago by research teams from the Czech Republic and Germany, who examined liquid and solid water clusters with the help of infrared spectroscopy. They concluded that the smallest snowflake of the world consists of ~ 475 water molecules (see ⬛ Fig. 1.22). Only when reaching this number of molecules, a transition occurs from loosely bound liquid-like aggregates without any regularity to a regularly periodically ordered solid ice crystal with hexagonal symmetry [15].

1.6 · The Hexagonal Symmetry of Snowflakes

◘ Fig. 1.20 Even snow crystal needles show a hexagonal symmetry: if we look from above or from the bottom, the hexagonal cross section of the inner wall will become visible. (Courtesy of Kenneth G. Libbrecht)

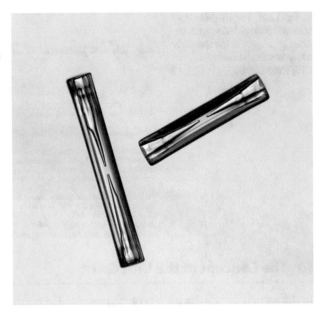

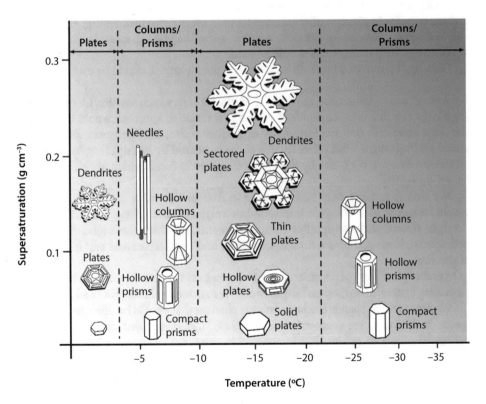

◘ Fig. 1.21 The form of snow crystals depends to a great extent on the temperature and humidity of the air layers in which they grow. The more humid, the larger and more branched the crystals are. (Adapted, courtesy of Kenneth G. Libbrecht)

1

Fig. 1.22 The article in the scientific journal *Science*, which gives an answer to the question of the smallest snowflake. ([15] screenshot)

21st of September 2012

A Fully Size-Resolved Perspective on the Crystallization of Water Clusters

Christoph C. Pradzynski,[1] Richard M. Forck,[1] Thomas Zeuch,[1*] Petr Slavíček,[2] Udo Buck[3*]

The number of water molecules needed to form the smallest ice crystals has proven challenging to pinpoint experimentally. This information would help to better understand the hydrogen-bonding interactions that account for the macroscopic properties of water. Here, we report infrared (IR) spectra of precisely size-selected $(H_2O)_n$ clusters, with n ranging from 85 to 475; sodium doping and associated IR excitation–modulated photoionization spectroscopy allowed the study of this previously intractable size domain. Spectral features indicating the onset of crystallization are first observed for $n = 275 \pm 25$; for $n = 475 \pm 25$, the well-known band of crystalline ice around 3200 cm^{-1} dominates the OH-stretching region. The applied method has the potential to push size-resolved IR spectroscopy of neutral clusters more broadly to the 100- to 1000-molecule range, in which many solvents start to manifest condensed phase properties.

1.7 The Concept of the Unit Cell

In this section we will discuss one of the most fundamental concepts in crystallography: the unit cell. For this purpose, we will look at some volume elements, abstract geometric bodies, and will consider in which smaller elements they can break down in order to give the original shapes if we assemble the smaller units together again. Thereby, two rules should be applied:

1. Only regular building units should be used.
2. Only one kind of building unit for a given volume element should be used.

Let us start with a large cube as shown in ▪ Fig. 1.23. In which smaller, identical building blocks can this cube be divided? Well, a first intuitive approach would be to divide the large cube into nothing but plenty of smaller cubes. This deconstruction process would end up with a mini cube. A cube is characterized by its edge lengths, which are all identical ($a = b = c$), and by the angles between its faces, which are all orthogonal ($\alpha = \beta = \gamma = 90°$).

But this is not the only possible way to dissect the cube. A further possibility would be to cut the cube into horizontal or vertical slices (▪ Fig. 1.24). This would result in square plates as smaller volume elements, in which the angles are again all orthogonal, however, now with a square basal and top surface and rectangular faces at the sides.

Another possibility of deconstruction is shown in ▪ Fig. 1.25. This results in rectangular prisms, again with orthogonal angles; however in this case the lengths of all edges differ.

We now want to change the large volume element that should be dissected in smaller pieces. Let's look at the geometric body in ▪ Fig. 1.26. This object can be divided into hexagonal pillars or prisms. But in contrast to the other examples, a hexagonal prism consists of 8 faces, not 6. Is it possible to dissect a hexagonal prism further into smaller elements, which are regular, consist of 6 faces, and still give the large volume element if we assemble them together? Principally, two possibilities are conceivable, which are shown in ▪ Fig. 1.27. In the first case, the hexagonal pillar is cut along the long axis in two identical halves, resulting in a pillar with an isosceles trapezoid as basal plane. The second possibility is to cut the hexagonal pillar into three identical pieces, resulting in smaller blocks with basal planes whose lengths of edges would be identical and enclose an angle of 120°.

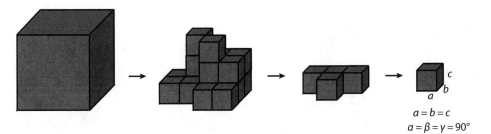

$$a = b = c$$
$$\alpha = \beta = \gamma = 90°$$

◘ Fig. 1.23 The deconstruction of a large cube into several identical small cubes

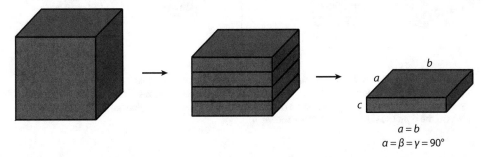

$$a = b$$
$$\alpha = \beta = \gamma = 90°$$

◘ Fig. 1.24 A further possibility to dissect a cube into identical building blocks, resulting in square plates

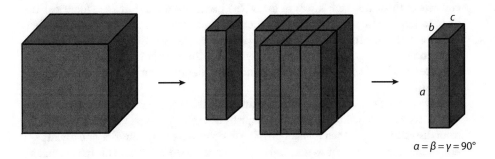

$$\alpha = \beta = \gamma = 90°$$

◘ Fig. 1.25 Breaking up of a cube into identical rectangular prisms

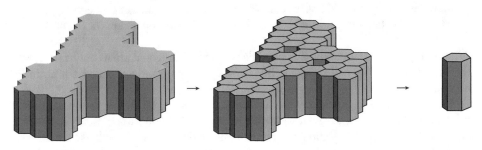

◘ Fig. 1.26 This geometric solid can be divided into hexagonal pillars

1

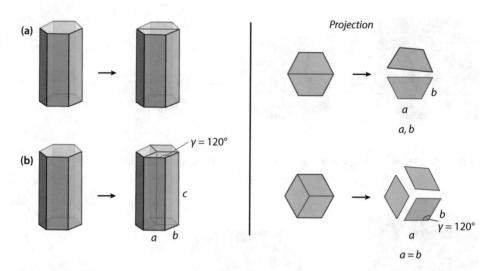

Fig. 1.27 Two ways of dividing hexagonal prisms (having 8 faces) into solids of only 6 faces: **(a)** Slicing along the middle of the long axis resulting in two pillars with trapezoid cross sections. **(b)** Dissecting from the center into three identical pillars with a rhombic basal face in which the edges enclose an angle of 120°. On the right the view from the top is shown

Which of these both possibilities should be preferred? To answer this question, we have to consider two aspects:

— The first way of dissection leads to a building block that is less regular, because it consists of three different kinds of faces (the basal and top surface are identical and lie parallel to each other, three of the four side faces are identical but having different orientations, while the fourth one has a different size). The second way leads to a geometric shape which consists only of two different faces (the bottom and top face are identical, and all four side faces are identical, each two of them being parallel).

— We have to keep in mind that we want to assemble the larger element out of the smaller ones. If we want to build up our initial shape by assembling the smaller ones resulting from dissecting way 1, then it is not possible to do so if we only consider one orientation of the smaller elements – necessarily gaps remain (■ Fig. 1.28a). This means that it is not possible to construct the large body or to fill the space completely by pure translation and assembling of the small element. It would be only possible, if we allow two different orientations. And this is exactly the difference to the result of dissecting way 2: here, it is possible to fill the space completely by joining the smaller elements together by this translation operation, even if we consider only one orientation (■ Fig. 1.28b).

Still it is a question, if and how it is possible to build a hexagonal prism out of the small element of way 2 – we will postpone this question for now and will address it again in ▶ Chap. 2.

We are now able to generalize the concept of the unit cell by finding an answer to the following question: Which regular geometric solids do fill the space completely by joining them together by pure translation operations along all three spatial directions? The answer is: These are parallelepipeds! "Epipedo" is Greek and means "face"; a parallelepiped is a geometric body confined by six parallelograms, of which two of each are congruent (super-

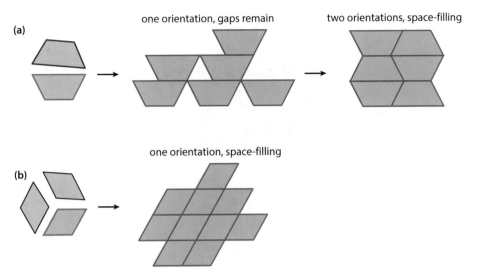

◻ **Fig. 1.28** (a) A trapezoid-like pillar does not fill the space completely, if only one orientation is allowed. (b) This is different to pillars with a rhombic face as basal plane, which indeed fills the space completely

imposable) and lie in parallel planes (◻ Fig. 1.29). In general, the lengths of the edges are different, and the angles between the edges are oblique. The most symmetric parallelepiped is a cube, at which all edges are of equal length and the angles between the edges are 90°.

Now we can define a unit cell of a crystal.

> The unit cell is the unit, which builds up the whole crystal by repeating translations (displacements) along all three spatial directions (◻ Fig. 1.30).

1.7.1 Characterizing the Unit Cell

1.7.1.1 Metric

The unit cell is, in its first instance, characterized by its so-called metric, meaning all of its lattice parameters (for the term lattice, see ▶ Sect. 1.9). We need exactly six parameters to unequivocally describe the unit cell. Three for the lengths of the edges, a, b, and c, which are running per definition along the x, y, and z axis, respectively, and three for the angles between the edges or faces, namely, the angle α, which defines the angle between edge b and c, the angle β that is enclosed by the edges a and c, and finally the angle γ, describing the aperture between edge a and b (◻ Fig. 1.31).

1.7.1.2 Symmetry

An additional feature of the unit cell follows directly from its definition and is related to its symmetry. The unit cell contains all symmetry elements of the crystal. This can be also expressed the other way round: If you choose a certain piece of a crystal as the unit

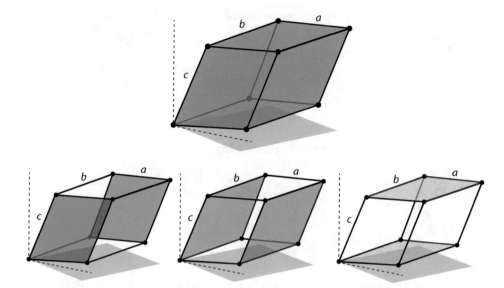

Fig. 1.29 A general parallelepiped, confined by six faces, two of which are congruent and lie in parallel planes

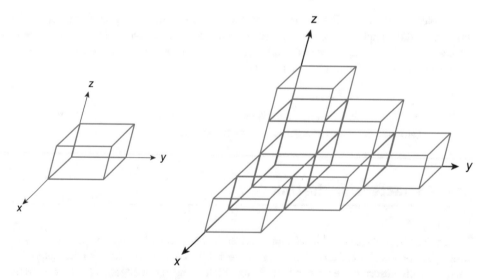

Fig. 1.30 By adjoining unit cells along all three directions of the system of coordinates, the whole crystal is built. A picture appears which is similar to that of a move, in which moving boxes are stacked – with the small difference that the crystals' boxes are sometimes rather skew and very tiny

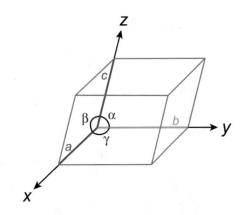

◘ **Fig. 1.31**   About the definition of the six lattice parameters

cell and apply the definition of the unit cell, this is we translate it along all three spatial directions and join them together, then no additional symmetry element can appear. If we chose a larger piece as unit cell and detect another symmetry element, which was not in the initially chosen smaller unit cell, then we would have chosen a unit cell being too small. And this in turn means that the symmetry aspect defines in a way the minimum size of a unit cell: the unit cell must just be so large that it contains all symmetry elements (apart from the translation as a whole). Later we will see that there is indeed a minimum size of the unit cell, but that it can be meaningful to choose a larger unit as the unit cell (▶ Sect. 2.3).

1.7.1.3 Chemical Composition (Stoichiometry)

Out of the definition of the unit cell, a further attribute follows, namely, regarding its chemical composition (the stoichiometry as the chemist say): the chemical composition of the unit cell must be identical to that of the whole crystal. Say, we have an ice crystal with the chemical composition H_2O. The ratio of the elements hydrogen (H) to oxygen (O) is then 2:1. And this must be also the ratio for the unit cell! If we had chosen a unit cell composed of the chemical species OH with a ratio H:O of 1:1, then also the whole crystal would necessarily have the same composition. This is because the only thing that is allowed is replicating and adjoining the unit cells along all three spatial directions – this would never result in a ratio of H:O of 2:1. Of course, it is possible that the unit cell contains more than one water molecule, for instance, 12 as in reality, but this does not change the element ratio.

1.8 The Seven Crystal Systems

In this section we want to classify unit cells. We have already seen some kinds of unit cells; they have in common that they are all parallelepipeds. Interestingly, the number of principally different unit cell shapes – expressed by their metric – is very limited. All crystals of the world can be classified according to not more than seven crystal systems. These

1

▣ Table 1.1 The seven crystal systems, their restrictions concerning the metric, and their maximal symmetry

Crystal system	Restrictions concerning the		Maximal symmetry
	Length of axes	Angles of the cell	
Triclinic	None[a]	None[a]	$\bar{1}$
Monoclinic	None[a]	$\alpha = \gamma = 90°$	$2/m$
Orthorhombic	None[a]	$\alpha = \beta = \gamma = 90°$	mmm
Tetragonal	$a = b$	$\alpha = \beta = \gamma = 90°$	$4/mmm$
Trigonal	$a = b$	$\alpha = \beta = 90°, \gamma = 120°$	$\bar{3}/m$
Hexagonal	$a = b$	$\alpha = \beta = 90°, \gamma = 120°$	$6/mmm$
Cubic	$a = b = c$	$\alpha = \beta = \gamma = 90°$	$m\bar{3}m$

[a]This means that the respective parameters can have any conceivable value

crystal systems have certain names, which you simply have to memorize. Roughly ordered according to rising symmetry they are called:
— Triclinic ("three times skew")
— Monoclinic ("one time skew")
— Orthorhombic ("everything is orthogonal")
— Tetragonal
— Trigonal
— Hexagonal
— Cubic ("dice")

These seven crystal systems differ with respect to their lattice parameters; however, the classification scheme is not based upon the metric itself, but rather upon their symmetry, which in turn leads to certain values for the lattice parameters. These interrelationships will become clearer soon. First of all, the seven crystal systems are gathered in ▣ Table 1.1 together with their specified metric and a symbol that represents their maximal symmetry. These symbols are hitherto unknown to you but will be explained in detail later.

For a better understanding let us shortly discuss the table rows.

In the triclinic crystal system, there are no restrictions concerning the lengths of the edges and also no restrictions with respect to the angles between them.

In the monoclinic crystal system, there are also no restrictions with regard to the lattice constants; however the degrees of freedom concerning the angles are somehow restricted, because two angles are determined to 90°, while the third one can have arbitrary values. You might ask, why theses fixed angles are chosen to be α and γ and not α and β or β and γ. The answer is simple: it is an agreement, nothing more. The monoclinic crystal system is characterized by exactly two orthogonal angles – in fact caused by symmetry reasons. The third angle can be (accidently) also 90° but does not have to be 90°. And in order to avoid that one crystallographer names the angle that not necessarily has to be 90° α and another one β or γ, the agreement was set that it should be always β.

In the orthorhombic system, the symmetry determines all angles to be orthogonal, while the lengths of the edges can be of any values and different to each other.

The next three crystal systems (tetragonal, trigonal and hexagonal) are characterized by the fact that two of the lattice constants have to be equal and that all three angles are fixed, whereby for the tetragonal systems, all angles have to be equal to 90°, and for the tri- and hexagonal, two of them have to be 90° (α and β) and the third 120° (γ).

The crystal system with the highest symmetry is the cubic one, in which all lattice parameters are fixed; there are no degrees of freedom left, all angles have to be 90° and the edge lengths all have to be same.

Although there are seven crystal systems, two of them show identical restrictions concerning their lattice parameters, namely, the tri- and hexagonal system. That is the reason why some people refer to only six crystal families. The tri- and hexagonal system are subsumed to the hexagonal crystal family, because their unit cell shapes are identical. However, the classification into the crystal systems is based upon symmetry, which is different for the tri- and hexagonal system as can be seen by the last column of ◻ Table 1.1. We will come back to this aspect in ▸ Sects. 2.7 and 2.8.

Regarding the metric sometimes also the term degrees of freedom are mentioned. This is another way to express what here is introduced as "restrictions." The number of degrees of freedom is given by the number of lattice parameters (always 6) minus the number of parameters, which are determined by symmetry. For the triclinic crystal system, there are six degrees of freedom, in the orthorhombic three, and in the cubic one zero,[1] respectively.

In virtually all books about crystallography you can find an overview of the crystal systems in the manner of ◻ Table 1.1. However, you should be aware of the fact that in most of the cases – a laudable exception is the book by W. Massa [16] – it is displayed in an incorrect or at least ambiguous manner. This is caused by the circumstance that not only restrictions are specified, but also statements concerning those lattice parameters are made, which are *not* explicitly determined by symmetry. These – attention: wrong! – tables look like the one in ◻ Table 1.2.

All entries of ◻ Table 1.2 printed in red are incorrect. For instance, take a look at the specifications of the triclinic crystal system: it is specified that all lengths and angles have to be different and that all angles have to be unequal 90°. It is, indeed, not very likely that crystals belonging to the triclinic system have unit cell lengths that are all identical and that all angles are equal or even equal to 90° but it is (mathematically) not forbidden. The symmetry of the triclinic system does indeed allow lattice parameters so that all unit cell lengths are identical; also the angles are not systematically forced to be different and unequal to 90°. Likewise, the angle β in the monoclinic case does not have to be unequal 90°, it can accidentally have a value of 90°.

> ❯ The lattice parameters only give an indication to which crystal system a given crystal probably could belong. It is not the metric that determines the symmetry, but rather the opposite is true: the symmetry determines certain values of the lattice parameters, although not in an unequivocal or biunique way.

---

1   To exclude misinterpretations: if there are no degrees of freedom left, then this does not mean that the unit cell lengths have a specific value, say 10 Å. It means that this value cannot be modified without necessarily altering the other values, too.

1

□ **Table 1.2** Unfortunately, common though wrong overview of the metric of the seven crystal systems

Crystal system	Cell constants	Angles
Triclinic	$a \neq b \neq c$	$\alpha \neq \beta \neq \gamma \neq 90°$
Monoclinic	$a \neq b \neq c$	$\alpha = \gamma = 90°, \beta \neq 90°$
Orthorhombic	$a \neq b \neq c$	$\alpha = \beta = \gamma = 90°$
Tetragonal	$a = b \neq c$	$\alpha = \beta = \gamma = 90°$
Trigonal	$a = b \neq c$	$\alpha = \beta = 90°, \gamma = 120°$
Hexagonal	$a = b \neq c$	$\alpha = \beta = 90°, \gamma = 120°$
Cubic	$a = b = c$	$\alpha = \beta = \gamma = 90°$

> Crystals that belong according to their symmetry to the triclinic crystal system might have accidentally lattice parameters of $a = b = c$ and $\alpha = \beta = \gamma = 90°$, while a cubic crystal necessarily, namely, out of symmetry reasons, must possess these lattice parameters.

In order to deepen our understanding of the relationship between the metric and the crystal systems, we want to go through the following three examples, in which from experimentally obtained lattice parameters possible crystal systems should be derived.

Example 1

A crystal structure analysis of a crystal sample gave the following lattice parameters:

$a = b = 12.0$ Å, $c = 12.9$ Å, $\alpha = \beta = \gamma = 90°$

To which crystal systems could the crystal belong to?

Solution: Because the cell constant c is different to a and b, we can exclude the cubic crystal system. The tri- and hexagonal crystal system also do not fit, because here the angle γ should be 120°. The requirements of the tetragonal system are fulfilled though: $a = b$ and all angles are 90°. This is also the case for the orthorhombic system. If the edge lengths can have arbitrarily values, then they could also adopt the specified values of our sample. And the angles are indeed, as the orthorhombic systems requires, all 90°. However, the values do fit also to the monoclinic and triclinic system. This means, we cannot conclude precisely to which crystal system this sample belongs. The given values match with the restrictions of the triclinic, monoclinic, orthorhombic, and tetragonal crystal system. We can identify to which of these crystal systems the crystal sample actually belongs only if we carry out an analysis of the symmetry or atomic structure, with the tetragonal system being the most probable.

Example 2

The lattice parameters have been determined to:

$a = b = c = 10.0$ Å, $\alpha = \beta = \gamma = 90°$

Solution: The tri- and hexagonal crystal system can be excluded because the angle γ is not 120°. However, the given lattice parameters are compatible with all other crystal systems. The crystal simple could belong to the triclinic, monoclinic, orthorhombic, and tetragonal as well as to the cubic system. The latter is, of course, the most probable one, but the others cannot be excluded a priori.

Example 3

The determination of another crystal gave the following lattice parameters:

$a = b = 10.0$ Å, $c = 12.0$ Å, $\alpha = \beta = 90°$, $\gamma = 120°$

Solution: The cubic, tetragonal, and orthorhombic crystal system require γ to be 90°. They can therefore be excluded. The tri- and hexagonal system, however, fit perfectly well. Because the lattice parameters can have arbitrary values for the triclinic system, it is one of the possible systems. The question is, if not also the monoclinic system could be a possible solution, because for this two of the angles have to be equal to 90°, which is obviously the case here. The only difference would be that the angle that is not restricted in any way is named differently, namely, γ. Indeed, the monoclinic crystal system would be a possible solution as well, if we only considered the symmetry: general symmetry considerations for the monoclinic crystal system lead to the requirement that two of the angles have to be equal to 90° and that the remaining third angle possibly deviates from that value. And we would indeed be free to name the angles that are fixed to 90° α and β, or α and γ, or β and γ. The only reason to call them α and γ is an agreement of the crystallographic community! In other words, in the monoclinic crystal system, symmetry elements can occur only in one (viewing) direction, and this is, by convention, the b direction, which runs by definition perpendicular to the (a,c) plane. And the angle between the a and c axis is, again by definition, the angle β. These definitions and conventions have been taken already into consideration during the analysis of the lattice parameters of the crystal sample; therefore the monoclinic crystal system can be excluded as well. The appearance of symmetry elements in certain directions of the crystal system will be treated intensively in ▸ Chaps. 3 and 4.

To conclude, possible crystal systems for the given sample of example 3 are the triclinic, the tri-, as well as the hexagonal system.

Finally, a kind of an empirical proof should be provided for the given statements, because you could ask: it might be possible that it is mathematically not forbidden that the angle β in the monoclinic crystal system has a value of 90°, however, how likely is it that a crystal exists, which is on the one hand relatively asymmetric so that it only belongs to the monoclinic crystal system, but on the other hand has an inner structure so that all angles of the unit cell are equal to 90°? The answer is that it is very unlikely, but that there are a few ascertained cases, in which this is indeed the case. One of these examples is the crystal structure of the compound sodium potassium zinc diphosphate [17], which crystallizes in the space group $P2_1/n$ and therefore belongs to the monoclinic crystal system. The lattice parameters are as follows: $a = 12.585(5)$ Å, $b = 7.277(5)$ Å, $c = 7.428(5)$ Å, $\alpha = 90.0$, $\beta = 90.0(5)°$, $\gamma = 90.0°$; the last digit in parentheses reflects the inaccuracy of measurement. Thus, the unit cell of this diphosphate compound has three orthogonal angles, but it does not belong to the orthorhombic crystal system. It has a very interesting channel-like structure, which is shown in ▪ Fig. 1.32.

1

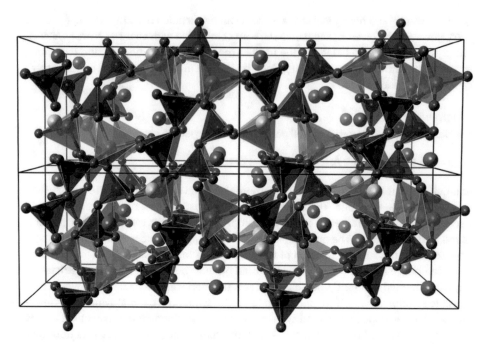

◘ **Fig. 1.32** Structure of $NaKZnP_2O_7$; view along the *b* axis; PO_4 (purple) and ZnO_4 tetrahedra (gray) form a channel-like structure, in which sodium (yellow) and potassium ions (blue) are incorporated

1.9 Lattice and Motif (Basis)

Now we want to come back to the term crystal structure, which was outlined in the forgoing sections but not explicitly explained.

There is a convenient short formula by which the term crystal structure is perfectly described: the crystal structure consists of a lattice plus the motif. The motif is also called basis. Let us first consider the term lattice. A lattice is a mathematical construct and can be defined as follows:

> A lattice is an infinite arrangement of mathematical points in space (3D) or in the plane (2D) or on a line (1D), in which all points have the same surrounding.

In ◘ Fig. 1.33a a section of a two-dimensional lattice is shown. A genuine lattice would be, of course, infinite. However, it is obvious that every point of the lattice has the same surrounding. In ◘ Fig. 1.33b three points and their surroundings are highlighted: point 1 (red) has 4 nearest neighbors (green), 8 next-nearest (blue), 12 next-next-nearest neighbors (orange), etc. The same is true for point 2 (and every other point). Now it is also clear, why a lattice – according to the given definition – has to be infinite extended: point 3 at the border of the section has a different surrounding compared to points in the center. Therefore, there must be no border. Real crystals, however, are not infinitely large and do have

□ Fig. 1.33 (a) A section of a two-dimensional lattice; **(b)** in a lattice all lattice points have the same surrounding; this is exemplified with the help of point 1 and 2; point 3 is at the border and has, obviously, not the same surrounding – therefore, mathematically, a lattice is in the proper meaning of the word infinitely extended

(a)

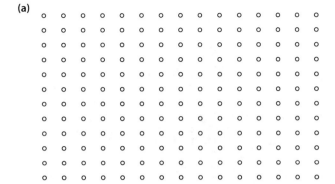

(b)

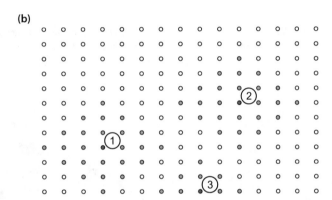

boundary surfaces. This has consequences for the chemical and physical properties at these interfaces, which are different to that of the bulk phase.

The question is now, what does such a lattice point represent? A possible interpretation is to consider it as a connection point between neighboring cells. In □ Fig. 1.34a a section of a crystal in the form of joined unit cells is shown. If we now placed a lattice point at every corner of the unit cells and removed the borders, we would obtain the "naked" lattice. Restricted to two spatial dimensions the result looks like as depicted in □ Fig. 1.34b.

A lattice is unequivocally characterized by its lattice vectors. These are such translation vectors, which transfers the lattice points into the other ones along the three directions a, b, and c. Occasionally, it is also said that these three lattice vectors – characterized by their lengths and directions – *span* the unit cell. In □ Figure 1.34b both the lattice vectors b and c of the two-dimensional lattice are shown. They run along the y and z axis, respectively. Lattice vectors can be considered as a kind of road signs: If I am standing at a particular lattice point, in which direction and which distance do I have to go to get to the next lattice point?

The term lattice should now be resolved, so let us continue with the motif. Until now unit cells were introduced as pure geometric envelope, as certain confined spatial figures (parallelepipeds). But the unit cells are, of course, not empty, they are filled with the motif/base. The motif consists of an arrangement of chemical species of the crystal. In real crys-

1

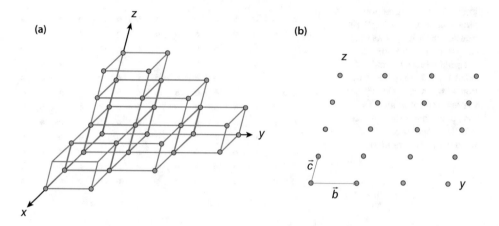

◻ **Fig. 1.34**   **(a)** Lattice points can be interpreted as connection points between unit cells. **(b)** Restriction to two dimensions and removal of the boundaries of the unit cells leads to the "naked" lattice

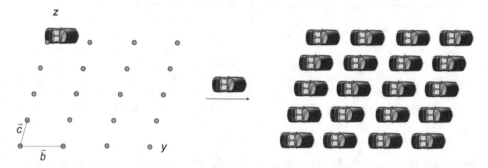

◻ **Fig. 1.35**   If a lattice point in this two-dimensional lattice represented a cabriolet, a parking lot-like car crystal would emerge

tals these are atoms or molecules; however, in principle a motif can be anything, for instance, also a car. Each (mathematical) lattice point represents a (real) motif. If we now apply the translation principal to the motif, a periodic, crystalline arrangement originates. In the case of the car a kind of parking place, with pure identical cars – a two-dimensional "car crystal" – see ◻ Fig. 1.35.

Because crystals are usually not made of cars, a chemically more realistic example is shown in ◻ Fig. 1.36. Here, the motif consists of a three-atomic molecule, and this leads to a respective molecular crystal. It can be inferred that all of the orange atoms form an orange lattice, the green atoms a green lattice, and the blue atoms a blue lattice (◻ Fig. 1.37). This means that all constituents of a crystal structure, all constituents of a motif are subject to the same translation principle. All of the translation lattices are parallel shifted with respect to each other, but they are all congruent (superimposable).

It is important to understand that the lattice is only a virtual construct that describes the distance and direction from one motif to another in all the spatial directions of the respective system of coordinates of the crystal system.

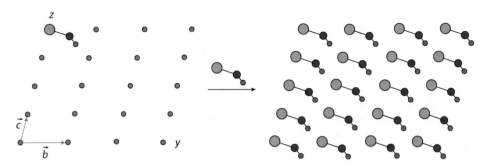

Fig. 1.36 Application of the translation principle on the motif of a three-atomic molecule

Fig. 1.37 All constituents of a motif give congruent (superimposable) translation lattices

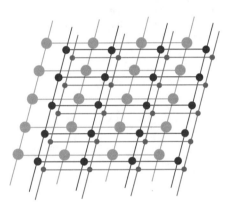

■■ Summary

In this chapter we have defined, what a crystal is, have discussed what features the term crystal structure comprises, and have derived, why crystals possess anisotropic properties. Furthermore, the correspondence principle was introduced. With the aid of two examples – quartz and snowflakes – it was shown how this principle manifests itself. After having introduced the concept and characteristics of unit cells, the seven crystal systems were discussed. Finally, the relationship between the term lattice and motif was clarified.

References

1. Eames Ch, Eames R (1977) Powers of ten. http://www.eamesoffice.com/the-work/powers-of-ten/. Accessed 28 Sept 2019
2. Rubin AE, Ma C (2017) Chem Erde-Geochem 77:325–385. https://doi.org/10.1016/j.chemer.2017.01.005
3. St. Lovgren (2007) Giant crystal cave's mystery solved. National Geographic, https://www.nationalgeographic. com/science/2007/04/giant-crystal-cave-mexico-mystery-solved/. Accessed 28th Sept 2019
4. Than K (2010) Giant crystal caves yield new "ice palace," More, National Geographic, https://news.nationalgeographic.com/news/2010/10/101007-lost-crystal-caves-mexico-science-mine-superman-ice-palace/. Accessed 28 Sept 2019
5. García-Ruiz JM, Villasuso R, Ayora C, Canals A, Otálora F (2007) Geology 35:327–330. https://doi.org/10.1130/G23393A.1
6. Otálora F, García-Ruiz J (2014) Chem Soc Rev 43:2013–2026. https://doi.org/10.1039/c3cs60320b

1

7. Bök C (2003) Crystallography. 2nd edn. Coach House Press, Toronto
8. Hooke R (1665) Micrographia: or some physiological descriptions of minute bodies made by magnifying glasses. J. Martyn and J. Allestry, London
9. Molčanov K, Stilinović V (2014) Angew Chem Int Ed 53:638–652. https://doi.org/10.1002/anie.201301319
10. The kilogram and the mole: counting atoms, Physikalisch-Technische Bundesanstalt (PTB), https://www.ptb.de/cms/en/research-development/research-on-the-new-si/ptb-experiment/the-kilogram-and-the-mole-counting-atoms.html. Accessed 28 Sept 2019
11. Kikuchi K, Kameda T, Higuchi K, Yamashita A (2013) Atmos Res 132–133:460–472. https://doi.org/10.1016/j.atmosres.2013.06.006
12. Libbrecht KG (2001) Eng Sci 64:10–19
13. Libbrecht KG (2005) Rep Prog Phys 68:855–895. https://doi.org/10.1088/0034-4885/68/4/R03
14. Furukawa Y, Wettlaufer JS (2007) Phys Today 60(12):70–71. https://doi.org/10.1063/1.2825081
15. Pradzynski CC, Forck RM, Zeuch T, Slavíček P, Buck U (2012) Science 337:1529–1532. https://doi.org/10.1126/science.1225468
16. Massa W (2004) Crystal structure determination, 2nd edn. Springer
17. Shepelev YF, Lapshin AE, Petrova MA (2006) J Struct Chem 47:1098–1102. https://doi.org/10.1007/s10947-006-0431-4

Crystal Shapes and Bravais Lattices

© Springer Nature Switzerland AG 2020
F. Hoffmann, *Introduction to Crystallography*, https://doi.org/10.1007/978-3-030-35110-6_2

2.1 Phenomenology of External Crystal Shapes

When you take a tour through a natural history museum with a mineral collection you will notice that crystals exist in an incredible variety of shapes. Almost no crystal looks like another with regard to its outer shape, that is, the *morphology* (Greek μορφή, morphé = form, shape). They range from thin, long, needlelike crystals to chunky blocks, bizarre-looking tufts grown together, and also flat, columnar, dandruff-like or cubic crystals can be found. In ◘ Fig. 2.1 a small selection of different crystals is shown.

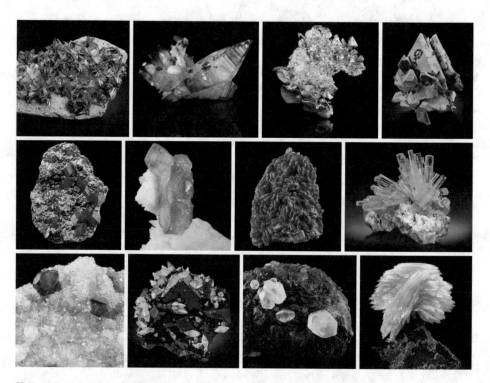

◘ **Fig. 2.1**   A tiny selection from the almost innumerable different crystal forms that can be found in nature. Line by line from left to right, the following exhibits of the Mineralogical Museum of Beirut are shown: rutile, collection number 1022, photographer FMI/James Elliott; corundum in the sapphire variety, collection number 1543, photographer AINU/Augustin de Valence; gold, collection number 1679, photographer AINU/Augustin de Valence; kesterite, collection number 1034, photographer FMI/James Elliott; rhodochrosite, collection number 584, photographer AINU/Augustin de Valence; baryte, collection number 1305, photographer AINU/Alessandro Clemenza; rhodonite, collection number 493, photographer AINU/Augustin de Valence; beryl in the aquamarine variety, collection number 1092, photographer FMI/James Elliott; scorodite, collection number 679, photographer AINU/Augustin de Valence; calcite on fluorite, collection number 1450, photographer FMI/James Elliott; sulfur, collection number 411, photographer AINU/Augustin de Valence; baryte, collection number 1549, photographer FMI/James Elliott. (Imprint by kind permission of © Mineralogical Museum Beirut. All rights reserved, ► www.mim.museum)

The question now is: How is this possible? In ▶ Chap. 1 we learned that crystals are constructed by only seven different building blocks and that a crystal is formed by assembling unit cells of the same kind. One reason for the existence of that many different crystal shapes is that in nature very often crystals are not found as single-crystals but as aggregates of a much larger number of small(er) single-crystals. And these aggregates were assembled – following the whim of nature, that is, the conditions of the environment at their time of formation – completely randomized. But even if we focus on single-crystals, the number of different shapes is very high (◘ Fig. 2.2). And not only that, there are many cases where even one and the same chemical substance forms crystals of different shapes. ◘ Figure 2.3 shows a photo of a beautiful specimen of intergrown pyrite crystals. Some of them seem to penetrate each other, but most important is the prevailing cubic shape. The cube is the most common shape of pyrite crystals. And that is not surprising because pyrite crystallizes indeed in the cubic crystal system. The interesting thing is that you will not only find cubic pyrite crystals but often also all the forms that are shown in ◘ Fig. 2.5. Besides the cubic shape, often octahedra, rhombic dodecahedra, pentagonal dodecahedra, and many more can be found. Altogether, more than 60(!) different shapes of pyrite crystal are known. (An informative article on the morphology of pyrite in sulfur-rich sediments can be found in [1].)

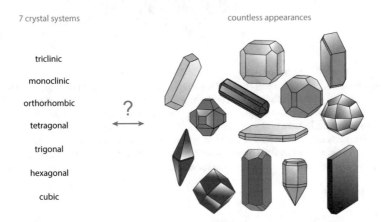

7 crystal systems

triclinic

monoclinic

orthorhombic

tetragonal

trigonal

hexagonal

cubic

?

countless appearances

◘ **Fig. 2.2** The seven crystal systems are juxtaposed with an incomparably larger number of crystal shapes

Chemistry and Structure of Pyrite

Chemically viewed, pyrite is iron(II) disulfide, and it belongs to the cubic crystal system with the space group $Pa\overline{3}$ (for the space group symbols, see ► Chap. 5). ◘ Figure 2.4 shows two views of the unit cell. The structure can be considered as a derivative of the well-known sodium chloride (rock salt) structure. Instead of the sodium ions, iron(II) ions (in red) occupy the corners and the center of the unit cell, and instead of the chloride now disulfide ions are located at the centers of the edges. An octahedral coordination sphere is formed around the iron ions (from six sulfur atoms) and a tetrahedral one around the sulfur atoms (from one sulfur and three iron atoms).

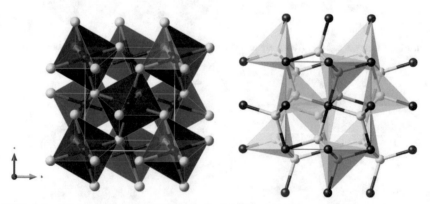

Fig. 2.4 Crystal structure of pyrite, iron(II) disulfide. The structure on the left shows the coordination environment of the iron ions, which are surrounded by octahedrons made of six sulfur atoms. The structure on the right highlights the coordination sphere of the sulfur atoms, which are surrounded in the form of a tetrahedron by three iron atoms and another sulfur atom

■ **Fig. 2.5** Some selected shapes of pyrite crystals. Overall, over 60 different shapes are known

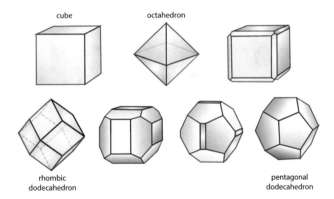

While looking at the shapes of ■ Fig. 2.5, you may have already noticed that some of them are elementary individual shapes, while others look as if they are composed of two different geometric bodies. That's indeed the case! But how can we explain the appearance of all these shapes, even though the unit cell of pyrite is cubic and the only way to build a macroscopic crystal is to join unit cells together?

Let us consider ■ Fig. 2.6, which shows two different ways of assembling small cubes: At the top, the relative growth velocity in all three spatial directions is absolutely identical. Necessarily, the outcome will be a large cube. But the speed of growth does not have to be the same for all spatial directions: at the bottom part of the figure, we see how adjoining small cubes results in a large octahedron at the end! This means, the secret of different outer crystal shapes, even within the same crystal system, is that the relative growth rate may vary for the different spatial directions. And that also means that the final crystals have different formation histories. The deeper reason that a crystal builds a particular shape is due to the prevailing conditions at the time of its growth, e.g., the number of crystallization seeds, the temperature, the pressure, the concentration of its constituents, whether these variables were constant during the growth process or not, and many more things.

Investigating the growth of crystals can quickly become very complicated and entire books have been written about this topic [2–4]. In the first place, it is important here to note the following: you may now be under the impression that crystals can grow in an arbitrary manner because the speed of growth in the different spatial directions can be different. However, that is not the case. If you classify them according to their symmetry, you will see that there are only 32 principally different geometric bodies, which can be mapped in a certain way on the 7 crystal systems. We have also seen in ▶ Chap. 1 that almost "arbitrarily" shaped snowflakes exist, but all of them have one thing in common: they all show a six-fold rotational symmetry! And in that sense, the shape of a snowflake and that of a pyrite crystal is anything but arbitrary. More about these 32 crystallographic symmetry classes can be found in ▶ Chap. 3!

2

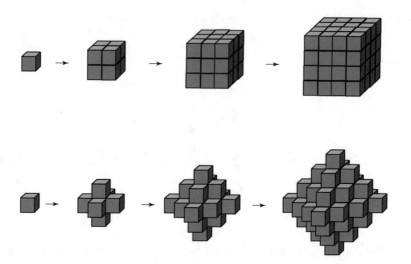

🔲 **Fig. 2.6**   Schematic representation of crystal growth starting from a cubic unit cell. At the top a large cube is formed out of small cubes, while in the bottom part, an octahedron is built

Crystal Shapes in 3D
A collection of visual math tools gathered at the Imaginary website (▶ https://imaginary.org/) is worth a visit. On the project page Math-to-Touch (▶ https://imaginary.github.io/applauncher2/), there is an interactive tool ("Polyhedra") that can be used to create very realistic, freely rotatable 3D crystal shapes resulting from a mix of various geometric basic bodies (Platonic solids). The sliders determine the size and shape of the facets. You will find it intuitively simple to operate this tool. 🔲 Figure 2.7 shows a screenshot.

🔲 **Fig. 2.7**   Screenshot of the web app "Polyhedra" at the Imaginary website that allows you to construct realistic 3D shapes of crystals by mixing geometric basic bodies

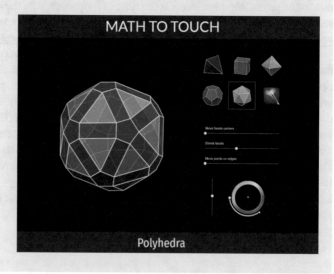

2.1.1 Tracht and Habitus

At this point, let us first focus on acquiring the necessary concepts for describing the morphology, the external shape of crystals. It is obvious that a description like our crystal at hand looks "somehow dodecagonal and thin" is inappropriate. Two terms are central to describe the morphology: *Tracht* (originally a German word for traditional costumes) and *habit*.[1] While the term Tracht describes the *number and combination of faces of the outer boundary of the crystal*, habit indicates the *relative size of the surfaces to each other*, from which the shape is built; these sizes have also an effect on the length to width ratio of the overall shape. While the terms for specifying the habit are very close to everyday life (plates, columns, needles, etc.), the Tracht describes precisely which type of faces occurs in which number. It describes, as the crystallographers say, which surfaces are *developed* on a crystal (meaning that they are visible). The term crystal habit is sometimes also used to describe not only single-crystals but the characteristic appearance of aggregates of crystals, such as spherulitic or botryoidal.

Two crystals can have the same Tracht, but a different habit, and the reverse is also possible: having the same habit, two crystals can develop different Trachts. This is best illustrated by an example shown in ◘ Fig. 2.8: all three crystals have the same Tracht, namely (from top to bottom), starting with a hexagon as the top surface, then six trapezoids appear as oblique boundary surfaces, followed by six rectangles as vertical-side surfaces and again six trapezoids; and another hexagon finally forms the base. This sequence of different types of faces is identical for all three crystals – they have the same Tracht. But, as you can see, the proportions of the faces are very different: for example, the rectangular faces of the crystal on the left side are very long and thin, while they are wide and flat for the crystal at the right. This results in a different habit, which is described by different, typical terms that reduce the form to the essentials: the crystal on the left can be described as platelike, the one in the middle as columnar and the right as block or as iso-metric (which means approximately uniformly extended in the three spatial directions).

In the second example (◘ Fig. 2.9), it should become clear that crystals with the same habit can have different Trachts. All crystals that are shown can be characterized as

◘ **Fig. 2.8**   Three crystals with the same Tracht (combination of faces), but different habit. (Redrawn from: Lapis, Chr. Weise-Verlag, Munich)

---

1   There is no equivalent term of "Tracht" in English; the same term "crystal habit" is applied to both. In this book we will still use both, because it is important to differentiate between both descriptors (see text).

isometric. But the cube is limited by 6 squares, the octahedron by 8 triangular faces, the rhombic dodecahedron – as the name implies – by 12 rhombi, etc.: these are all different "crystal dresses" of the same habit.

The terms Tracht and habitus are not only used in the field of mineralogy or crystallography but have also a social relevance. We can utilize this everyday use in order to memorize the concepts. In the left photo of ◘ Fig. 2.10, you can see my fellow researchers Carl and Tamás, who are almost equally dressed, like typical chemists – they have the same "costume." But while Carl is long and thin and therefore rather of the slender type

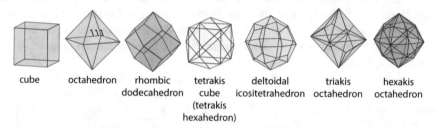

| cube | octahedron | rhombic dodecahedron | tetrakis cube (tetrakis hexahedron) | deltoidal icositetrahedron | triakis octahedron | hexakis octahedron |

◘ **Fig. 2.9**   All crystal shown here have approximately an isometric habit, but they appear in different "crystal dresses," Trachts: the number and type of the outer surfaces differs. As an example, on one of the triangular faces of the octahedron, the numerical sequence "111" is given, which precisely describes this face in relation to the internal structure of the crystal. It is a so-called Miller index (see ▶ Sect. 2.2)

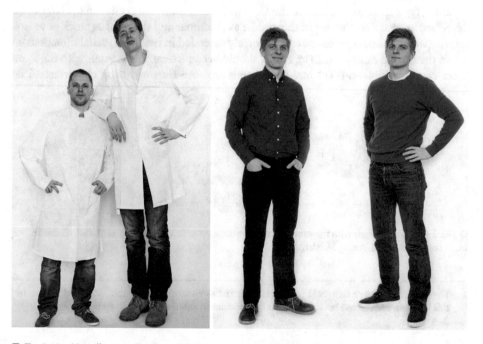

◘ **Fig. 2.10**   My colleagues Tamás and Carl in the same Tracht, but with distinctly different habitus (left) and my colleague Michael in two different Trachts, i.e., dresses (right)

("bean-pole"), Tamás impresses by pronounced mountains of muscles and is rather of the stocky type – different habitus. In the right side of ◘ Fig. 2.10, you can see two photographs of my colleague Michael, who of course always has one and the same habitus, but in the right photograph he looks rather sporty, while on the left picture, he is dressed a little more formal, and he looks more friendly; maybe he will go out for dinner with his wife... – he wears different Trachts.

In the next section (▶ Sect. 2.2), we want to learn how the outer surfaces of a crystal can be described in more detail and how they are related to their internal structure.

2.2 Miller Indices

In ◘ Fig. 2.9 (▶ Sect. 2.1.1) we have seen that three numbers (a triplet) are used for the exact denomination of crystal faces. These are the so-called Miller indices, which are used to denominate the crystal faces in a systematic way. Miller indices are also used to indicate lattice planes in the crystal lattice. At first glance, they seem a little complicated, but on closer inspection they are relatively easy to deduce, and, most importantly, they are very useful.

There was a series of scientific groundwork that finally led to the Miller indices. The Danish physician Nicolas Steno recognized in 1669 that the interfacial angles between corresponding faces on crystals are the same for all specimens of the same mineral – regardless of their size (*Steno's law of constant angles*). Steno first studied the prismatic faces of quartz crystals and then suggested that his observation could be a general property of all minerals. In 1783 Jean-Baptiste Romé de L'Isle published a detailed description of 500 crystal species and was able to empirically confirm the law of angular constancy.

But what does this law of constant angles actually mean? Let us look at ◘ Fig. 2.11: the interfacial angle between two adjacent face normals is always 60° for a hexagonal crystal, or in other words, two adjoining outer planes always form an angle of 120°. This angular constancy between analogous faces is valid for regular hexagonal crystals (red and green) as well as for one in which three of the six faces are less well or are better developed (blue) or even in "distorted" variants (pink and orange).

◘ **Fig. 2.11** Illustration of the law of constancy of interfacial angles (see text)

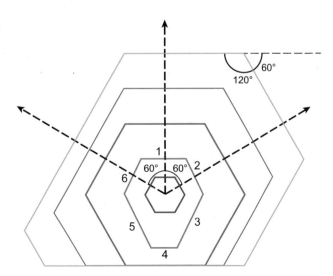

2

☐ Fig. 2.12 A blue specimen of the mineral haüyne (pronunciation [ha'ɥi: n], sometimes also named hauynite), which belongs to the class of aluminosilicates; crystal size, approx. 1 mm; locality, Rhineland-Palatinate, Germany

☐ Fig. 2.13 A two-dimensional lattice in which a certain number of lattice lines are drawn. The individual lattice lines of a given set are always oriented parallel to each other and have the same distance from each other. The unit cell is shown as a black rectangle

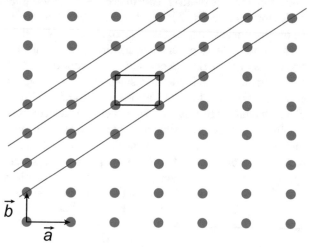

These findings have been further developed, i.e., by Christian Samuel Weiss (the founder of geometric crystallography) and René-Just Haüy, who is considered one of the fathers of modern crystallography – a mineral was also named after him; a beautiful specimen is shown in ☐ Fig. 2.12.

The ideas thus developed were finally taken up by William Hallowes Miller (British mineralogist and crystallographer), who introduced the system for unambiguous denomination of crystal faces, which was named after him [5]. To explain these Miller indices, it is useful to first introduce the notion of lattice planes. Lattice planes are a set of parallel planes of equal spacing intersecting the lattice in a specific manner. In ☐ Fig. 2.13 a 2D crystal lattice is shown in which a specific set of parallel lines (in 3D these would be planes) are drawn, all of which have the same distance to each other.

It is important to realize that these lines or planes are not physical entities. They are not made of stone or wood, but only abstract mathematical entities that connect lattice points. In principle, there are an infinite number of such lattice planes that pass through a crystal. In ☐ Fig. 2.14 two further set of lattice planes are plotted, in which the individual, parallel planes of

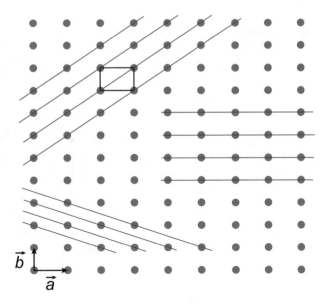

Fig. 2.14 In principle, there are an infinite number of lattice planes that run through a crystal; here, three possible sets of lattice planes are drawn

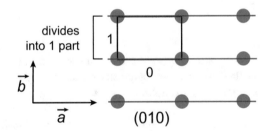

Fig. 2.15 Illustration for the derivation of the Miller indices of sets of lattice lines/planes

a set have the same distance from each other. When we are talking about a set of lattice planes, it does not mean a particular plane, but all planes together that are oriented in an identical way.

Now, Miller indices form a notation system for such lattice planes. They consist of three integer and coprime numbers h, k, and l, which is why they are also occasionally called hkl values. They are put in parentheses. We can determine them relatively simply by answering the following question: In how many parts do the planes of a set of lattice planes divide the lattice constants a, b, and c of the unit cell? For illustration purposes, let us consider two examples.

Example 1

The planes or lines in ◘ Fig. 2.15 do not intersect the axis a at all (they run parallel to it), the b axis is divided into exactly one part, and the c axis is, again, not cut, because here we consider only a 2D example. This results in Miller indices of (010), spoken as "zero-one-zero." Whenever an index is zero, it means that the planes do not intersect the corresponding axis or vice versa expressed: the intercept lies at "infinite."

Example 2

How would the set of lattice planes shown in ◘ Fig. 2.16 be named in the system of Miller?

2

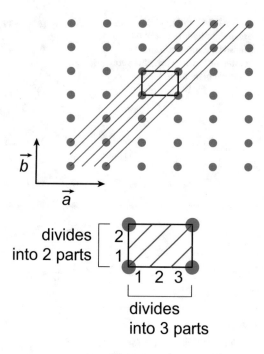

■ **Fig. 2.16** Another illustration for the derivation of the Miller indices of sets of lattice lines/planes

The lattice constant a is subdivided into three parts, the b axis into two and – as we again consider a two-dimensional example – the c axis is not cut. The resulting Miller indices are (320), that is, "three-two-zero."

Based on the examples, the following relationships can be derived with regard to the Miller indices, which are illustrated in ■ Fig. 2.17:

- The larger the indices, the smaller the distance d between two directly adjacent planes of a set of lattice planes; these distances are also called d values.
- The larger the indices, the smaller the density of lattice points (i.e., the number of lattice points per length unit) that lie on the corresponding planes. Since lattice points represent chemical motifs (hence atoms or atomic groups), this means, for example, that the density of atoms along faces with high Miller indices is relatively low, while faces with very low Miller indices are more densely packed. This also has consequences for the chemical reactivity, which can be very different for the different faces developed on a single-crystal.

Miller and Laue Indices

To be distinguished from the Miller indices are the Laue indices used in the X-ray or electron diffraction. Unlike Miller indices they do not have to be coprime, meaning that also indices like (220) or (422) occur. These indices refer to so-called higher diffraction orders, where the order n is the multiplication factor of the underlying set of lattice planes. In the examples given, the order $n = 2$ is the same in both cases, and the associated lattice planes are (110) and (211), respectively.

◻ Fig. 2.17 With increasing Miller indices, the distances between adjacent lattice planes become smaller and the density of lattice points decreases

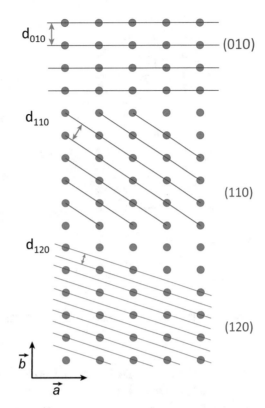

2.2.1 Negative Miller Indices

You may have already seen Miller indices with a dash or bar above the number. As in other crystallographic contexts, it means "minus" (it is either actually spoken as "minus" or as "bar"), the Miller index then has a negative sign. Let us consider ◻ Fig. 2.18, which shows two sets of lattice planes, the set (120) in the lower part and the set (1$\bar{2}$0) in the upper part ("one-minus two-zero"). Apparently, these sets of planes have the same gradient; they enclose the same angle with the a axis, but with opposite sign. In order to determine this, we will use the following procedure:

1. We first identify the plane of a set of planes that runs through the origin of the lattice. Of course, in order to do so, the coordinate system must be defined.
2. Now we take a plane directly adjacent to this identified origin plane, which lies in the direction of the positive plane normal and read off the intercepts. If these lie on the negative side of the coordinate system, the corresponding indices are given a negative sign.
3. Note that we always have to proceed in the same way for all the sets of lattice planes. In the second step (2.), we could have chosen also the neighboring plane, which lies in the direction of the negative normal of the plane. However, we would then have to proceed in the same way for all other lattice planes, too.

Miller indices in which the signs are inverted for all three values (h, k, and l), for example, ($3\bar{2}\bar{1}$) and ($\bar{3}21$), therefore denote the same set of lattice planes, i.e., it holds $(hkl) = -1 \times (hkl)$.

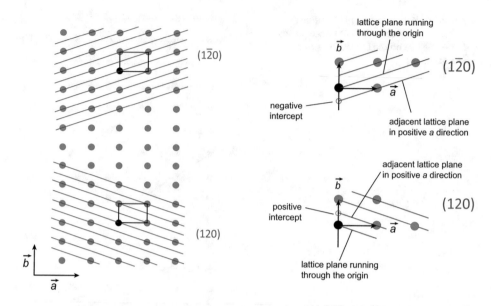

Fig. 2.18   Illustration for determining the sign of Miller indices (see text)

2.2.1.1 Exercise to Determine the Signs of Miller Indices of Lattice Planes

In ☐ Fig. 2.19 five sets of lattice planes are shown. Since there are two ways to choose the adjacent plane of the plane passing through the origin, there are two possible solution sets, which are shown in either the left or right box next to the drawing. It is important that you always act consistently, meaning the solution sets must not be mixed. Try to understand the given solutions!

2.2.2 Lattice Planes and Miller Indices in 3D

So far, we have considered only lattice planes or, to be precise, lattice lines of 2D lattices, but of course, crystals are three-dimensionally extended objects. To give an impression of how lattice planes look in 3D, in ☐ Fig. 2.20, three sets of lattice planes of one and the same lattice are shown: (100) in blue, (010) in green, and (111) in red.

2.2.3 Miller Indices, Weiss Parameters, Crystal Faces, and the Correspondence Principle II

The question now is how Miller indices, which we got to know as labels of lattice planes, and crystal faces relate to each other. This is relatively simple: the *outermost* plane of a set of lattice planes forms exactly the boundary face of a crystal. And that is the reason why

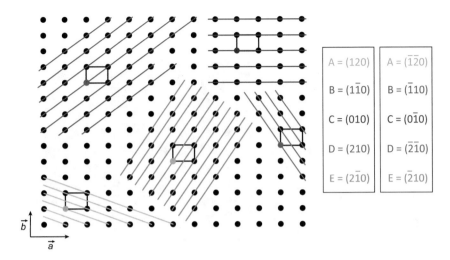

The table within the figure:

A = (120)	A = ($\bar{1}\bar{2}0$)
B = ($1\bar{1}0$)	B = ($\bar{1}10$)
C = (010)	C = ($0\bar{1}0$)
D = (210)	D = ($\bar{2}\bar{1}0$)
E = ($2\bar{1}0$)	E = ($\bar{2}10$)

Fig. 2.19 Exercise for determining Miller indices

the faces of crystals are denominated just like the corresponding lattice planes! This relationship confirms the correspondence principle (correspondence between the morphology and the structure of the crystal), which has already been dealt with in ▶ Sect. 1.5. Since Miller indices represent only relative parameters, a parallel shift (to the outermost edge of the crystal) does not change the relationships of the intercepts. What is also evident is that faces that are linked by the relationship $(hkl) = -1 \times (hkl)$ belong to the same set of lattice planes! This is illustrated exemplarily for the faces (110) and ($\bar{1}\,\bar{1}0$) by the thin red lines in the interior of the crystal of ◻ Fig. 2.21.

An open question, however, is how we index crystal faces when we have no knowledge of the crystal lattice and thus orientation of the lattice planes. In this case, the following rule applies: the *largest* surface, which cuts (if necessary in its extension) all three crystallographic axes, gets the index (111). This face is called the unit face. All other faces are then determined according to the relative intersection ratios to this unit face. These intercept ratios are also called *Weiss parameters*. Let us best consider an example (◻ Fig. 2.22):

In this example there are two different face types that intersect all three axes. One of the faces of these largest faces is marked in red in ◻ Fig. 2.22a. Thus, this face, which intersects all axes in the (defined) positive direction of the coordinate system, gets the index (111), and we define the intersections of the face with the respective axes as unit length along the axes $+a$, $+b$, and $+c$. The blue face in ◻ Fig. 2.22b also cuts all three axes and is smaller. Which index does this face get? Well, in relation to the unit face, which had the intercepts $1a$, $1b$, and $1c$, the intercepts are now $2a$, $2b$, and $2/3\,c$. We first build the reciprocals (1/2, 1/2, 3/2) and then convert them into integer, coprime parameters: this yield (113). In ◻ Fig. 2.23 all faces of the crystal are denominated with their corresponding Miller indices.

2

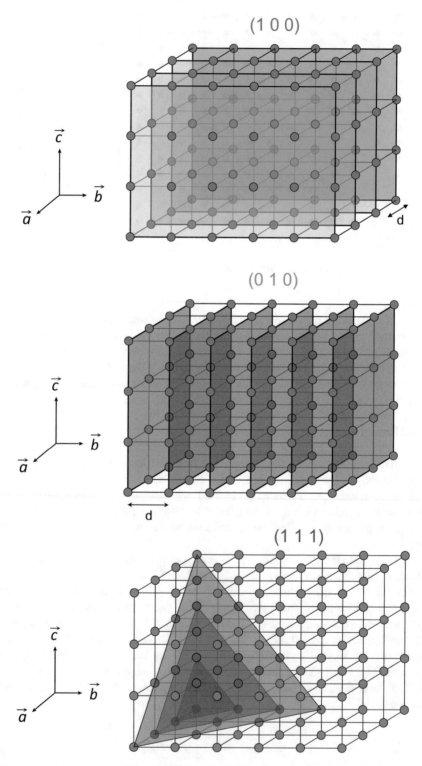

□ **Fig. 2.20**   The lattice planes (100), (010), and (111) of a 3D lattice

Fig. 2.21 The outermost lattice planes of a set of planes form the boundary faces of a crystal and are accordingly labeled with the same Miller indices as the lattice planes themselves

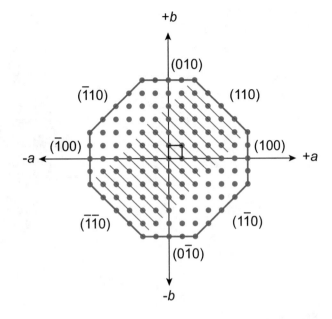

Fig. 2.22 (a) The largest face that intersects all three axes (where necessary in their extension, here in gray) in the positive direction is the unit face and gets the Miller indices (111), which are now called form symbol. (b) After this first determination, all other smaller faces can be easily indexed by their relative intersections with the axes in relation to the unit face. The blue face gets the indices (113); see text. The intersections with the axes are highlighted in yellow circles

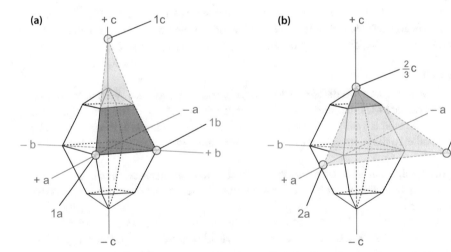

Three more examples of conversion of intercept ratios into Miller indices will be given below.

Example 1

The intercepts of a face should be at $-2a$, $3b$, and $4c$. What is the face symbol? We first build the reciprocals $(-1/2, 1/3, 1/4)$ and then form integer, coprime parameters. It follows $(\bar{6}43)$.

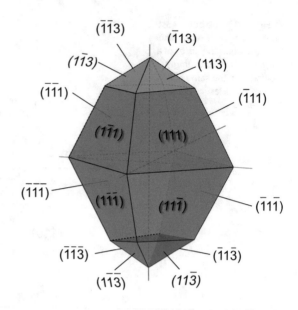

Fig. 2.23 Complete face indexing of the crystal shown in **Fig. 2.22**

Example 2

A crystal face has the intercepts $3a$, $6b$, and $8c$. What are its Miller indices? The reciprocals give 1/3, 1/6, and 1/8. Multiplication with the lowest common multiple of the numbers in the denominator (= 24) and canceling leads finally to the Miller indices (843).

Example 3

The intercepts of a face are determined as $-4a$, $6b$, and $8c$. Which face symbol do we get? The reciprocals are $-1/4$, 1/6, and 1/8. The lowest common multiple of the denominators is again 24. The result is $(\bar{6}43)$. The result is identical to that of Example 1; the face was only shifted parallel to that determined in Example 1 with the intercepts $-2a$, $3b$, and $4c$.

Now you should be able to correctly interpret face labels, for instance, such as in **Fig. 2.8. In ▶ Sect. 3.3.5, we will learn how to draw crystal shapes ourselves using a free software called VESTA!

To conclude this section, it should already be mentioned here as a small anticipation that equivalent faces – i.e., those that are symmetrically related – are grouped together and their hkl values (which are the same in absolute terms, i.e., except from their signs) are put in curly brackets: $\{h\,k\,l\}$. This is called a face symbol.

2.3 Bravais Lattices (Centerings) and the Choice of Unit Cells

2.3.1 Primitive Unit Cells/Lattices

In ▶ Sect. 1.8, the seven crystal systems were introduced. These seven crystal systems include *six* corresponding unit cell forms because the tri- and hexagonal crystal systems have the same metric (see **Table 1.1 in ▶ Chap. 1). The relationship between the unit cell and the corresponding lattice was established in ▶ Sect. 1.9, in which it was explained that the lattice points can be thought of as points of attachment of the unit cells, where eight unit cells adjoin at every corner. If we consider only such unit cells, in which lattice points are only at the corners, then we speak of primitive unit cells or lattices. The trigonal crystal system is special here, because there are so to speak two sorts of trigonal crystals: If we want to describe all of them with primitive unit cells only, then we need two different unit cell shapes. Some of the crystals of the trigonal crystal system can be described with the same unit cell shape as in the hexagonal crystal system, namely, a hexagonal prism. However, there are also crystals of the trigonal crystal system, which must be described in their primitive lattice form as a *rhombohedron*. The restrictions concerning the metric are here: $a = b = c$ and $\alpha = \beta = \gamma$. The primitive trigonal unit cell, however, can always be converted into a cell – by a so-called rhombohedral centering – which looks like the hexagonal and have, hence, the same metric ($a = b$, $\alpha = \beta = 90°$ and $\gamma = 120°$). This means that although there is a rhombohedral lattice, there is *no* rhombohedral crystal system. These relationships are explained in detail in ▶ Sect. 2.8.

The seven primitive unit cells or lattice types of the seven crystal systems are shown in ◻ Fig. 2.24. The lattice types shown in ◻ Fig. 2.24 are called primitive because they are indeed simple with respect to the distribution of lattice points: lattice points are present only at the corners of the unit cell, but not in the center of the unit cell nor on the edges or center of the faces. This means automatically that a primitive unit cell comprises exactly one base or motif (note: one base, not necessarily just one atom). It also means that the primitive unit cell of a given crystal system always automatically represents the smallest unit cell; a primitive unit cell can always be set up or found. The question, however, is

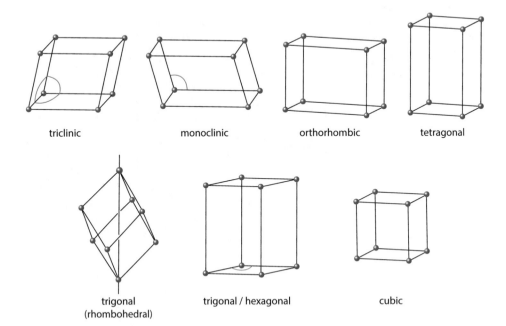

triclinic   monoclinic   orthorhombic   tetragonal

trigonal   trigonal / hexagonal   cubic
(rhombohedral)

◻ **Fig. 2.24** The seven primitive unit cells or lattice types of the seven crystal systems. In the primitive form, there are two different axis systems for crystals of the trigonal crystal system, a rhombohedral one with restrictions on the metric of $a = b = c$ and $\alpha = \beta = \gamma$ (typically but not necessarily the angles are $\neq 90°$) and a hexagonal one with the metric $a = b$, $\alpha = \beta = 90°$, and $\gamma = 120°$

2

whether the smallest unit cell is always the "best choice," with "best choice" meaning whether the smallest unit cell reflects the underlying symmetry in the best way. It turns out that this is not the case. This circumstance leads to the so-called centerings or to the 14 Bravais lattices.

In the following, we will discuss the question of how unit cells should be chosen in a sensible manner, and which rules should be followed. To get you in the mood for identifying point patterns and classifying unit cells, let me introduce you to the graph shown in ◘ Fig. 2.25, a so-called Marroquin pattern. Look at it for a while. What did you notice? Our brain has the permanent impulse to look for and to identify patterns. A selection of what you may have perceived as concrete patterns is superimposed on the graph in ◘ Fig. 2.26: square patterns, face-like elliptical patterns, three-point patterns, larger circular arrangements, and patterns with 12-fold radial symmetry. But you probably have noticed many more patterns. The interesting thing about this graphic or our perception of it is that because of the urge of our brain to constantly look for patterns, no pattern can be permanently fixed; in search of new patterns, those once discovered fade to the extent that a new one can be identified. In this graph this is the case every three seconds on average. It originates from the book *Vision: A Computational Investigation into the Human Representation and Processing of Visual Information* by David Marr [6]. David Marr is one of the founders of neuroinformatic, was a professor at MIT, and has worked extensively on the human perception and processing of visual information.

◘ **Fig. 2.25** Marroquin pattern; a computer graphic that illustrates the connection between visual perception and pattern recognition. (Courtesy of Dr. José L. Marroquin, Human visual perception of structure., M. Sc. Thesis, 1976)

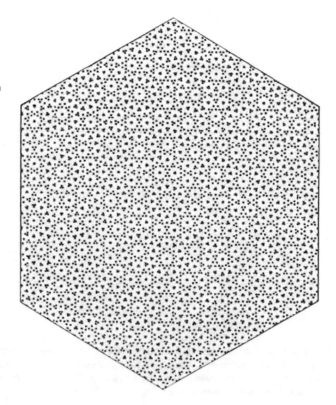

Fig. 2.26 Marroquin pattern with some outlined shapes that are temporarily perceived, but not permanently fixed

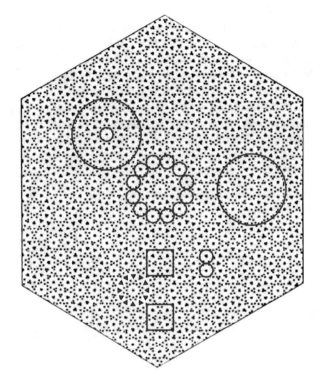

Fig. 2.27 An arrangement of lattice points. Which unit cell would you choose?

After this little digression, let us return to another kind of point patterns, viz., arrangements of lattice points to which the most meaningful unit cells are to be assigned. In ◘ Fig. 2.27 a certain arrangement of lattice points can be seen. Which unit cell would you choose?

2

It is not unlikely that you chose the unit cell drawn in black (■ Fig. 2.28), a primitive unit cell with lattice points only at the corners of the cell. An equivalent unit cell is drawn in red, which differs from the black only in that the origin was chosen differently. This is also a possible unit cell because it satisfies the condition of being able to generate the entire lattice by pure translation operations. If you actually chose the black or red unit cell, then you intuitively have chosen a unit cell appropriate to the point pattern.

There are other possible unit cells that are shown in green in addition to the previous unit cells in ■ Fig. 2.29; they are equivalent to the red or black unit cell; nevertheless, due to the following rules for selecting unit cells, the black or red unit cell is preferable:

■ **Fig. 2.28** Two possible and equivalent unit cells

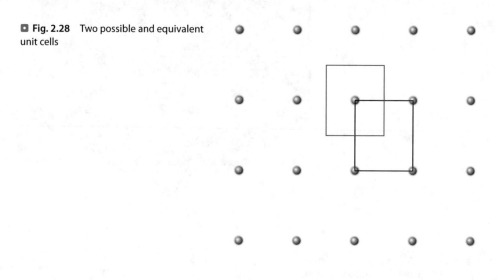

■ **Fig. 2.29** Of all the possible unit cells, the unit cells drawn in red and in black are to be preferred because their lattice vectors run perpendicular to mirror planes (dashed black lines, denoted by *m*)

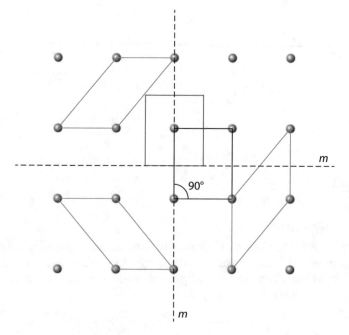

1. The unit cell should be as small as possible. This means that the lattice vectors should be as short as possible.
2. At the same time, the unit cell should reflect the symmetry of the lattice. This means that the lattice vectors should run parallel to symmetry axes or perpendicular to planes of symmetry.

   This is indeed the case for the black and red unit cells, where the lattice vectors are perpendicular to the symmetry planes marked with *m* in ◘ Fig. 2.29 (mirror planes and other symmetry elements are discussed in detail in ▶ Chap. 3), but not for the green unit cells.
3. An additional rule is that the lattice vectors should be orthogonal (or hexagonal), i.e., they should enclose an angle of 90° (or 120°). This is also the case for the black and red unit cell.

2.3.2 Centered Unit Cells

Let us consider another arrangement of lattice points in ◘ Fig. 2.30. Which unit cells could you choose here?

In ◘ Fig. 2.31 some possible primitive unit cells are shown. As you will quickly see, whatever primitive unit cell we choose, they are all oblique-angled – however, according to Rule 3, it is advisable to find unit cells where the lattice vectors are oriented orthogonally, if possible. What could a unit cell look like that complies with this rule? Well, the solution is shown in ◘ Fig. 2.32. Instead of the smallest possible unit cell, a primitive cell, a larger, centered unit cell was chosen, with an additional lattice point in the center of the cell. The choice of this unit cell now has the advantage that the lattice or the crystal can be described in a higher metric symmetry system, which means nothing more than that a larger number of angles is rectangular (or hexagonal).

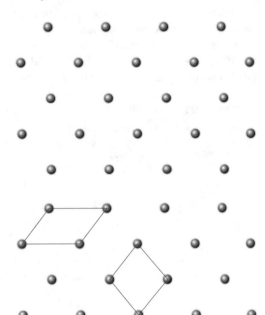

◘ **Fig. 2.30**　Another arrangement of lattice points. Which unit cell should be chosen here?

◘ **Fig. 2.31**　All primitive unit cells of this lattice are skewed

2

Fig. 2.32 In addition to the primitive unit cells (red, 1, and blue, 2), here a centered unit cell (green, 3) is shown, which contains an additional lattice point in the center of the cell

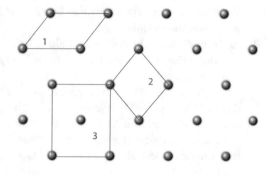

Fig. 2.33 If, for a given point pattern, the symmetry of the lattice is better manifested by the choice of a larger, centered cell and a larger number of right angles (or angles of 120°) are present, then this choice is preferable. By choosing a larger, centered unit cell, the symmetry of the lattice does not change – the two mirror planes (m) illustrate this because they are present both in the primitive cell 2 and in the centered cell 3. The choice of the primitive unit cell 1, however, would be particularly unfortunate because it does not reflect the symmetry of the lattice at all

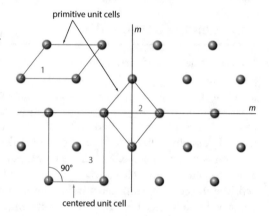

It does not mean that the choice of unit cell can change the symmetry or the crystal system of the lattice, because these are given intrinsically! And that, in turn, means that the two unit cells – the primitive and the centered – must have the same symmetry, except that it appears more obvious in the centered one, because here the lattice vectors are perpendicular to the mirror planes. These relationships are illustrated in ◻ Fig. 2.33.

Insight into the Transformation of Unit Cells: Reduced Cells
In ◻ Fig. 2.33 we saw several possible unit cells of the same lattice. Imagine that a colleague of yours ignored the convention of setting up unit cells and chose the unit cell marked with 1 or 2, which is certainly not forbidden! But you have chosen the unit cell that corresponds to the convention, i.e., that cell marked with 3. If you both now compare your crystal structures, you have a problem, because even the length and direction of their lattice vectors are different, and you wonder if you actually speak of the same crystal structure, where even the metric does not match.

Fortunately, there is a way to find out if the lattices are identical: they are identical if the cells can be converted into each other using a linear combination of the respective lattice vectors. What does that mean? Linear combination means that we can add or subtract the given vectors or multiples or fractions of them if necessary. And if we can use such a linear combination to convert the lattice vectors of one unit cell into the lattice vectors of the other unit cell, then the lattices are identical.

In the general form, a linear combination of two vectors looks like this:

$$\vec{c} = x \cdot \vec{a} \pm y \cdot \vec{b}$$

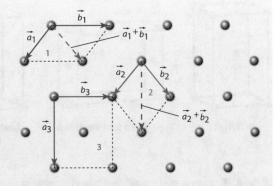

Fig. 2.34 Illustration concerning the transformation of unit cells, here from cell 1 to cell 2 and from cell 2 to cell 3

The result of such a vector addition can also be obtained graphically by placing all the arrows next to one another, taking their sign into account, and then drawing a new arrow from the origin to the end point of the attached arrows. Now we return to the present lattice of ◻ Fig. 2.33: how can the lattice vectors $\vec{a}$ and $\vec{b}$ of cell 1 be combined so that the vectors of cell 2 result? And how can the vectors of cell 2 be used to construct those of cell 3? The solution is shown in ◻ Fig. 2.34 and reads:

$$\vec{a}_2 = \vec{a}_1 \text{ and } \vec{b}_2 = \vec{a}_1 + \vec{b}_2 \text{ as well as}$$
$$\vec{a}_3 = \vec{a}_2 + \vec{b}_2 \text{ and } \vec{b}_3 = -\vec{a}_2 + \vec{b}_2$$

Obviously, the $\vec{a}$ vectors of cells 1 and 2 are identical, and the $\vec{b}$ vector of cell 2 is nothing more than the sum of the $\vec{a}$ and $\vec{b}$ vector of cell 1. The $\vec{a}$ vector of cell 3 is given by the sum of $\vec{a}$ and $\vec{b}$ vector of cell 2, and the $\vec{b}$ vector of cell 3 is obtained by the $\vec{a}$ vector in the negative direction plus the $\vec{b}$ vector of cell 2.

Now it could still be that the coordinate system was chosen differently, for instance, in that way that the principal choice of the cell was identical, but the lattice vectors $\vec{a}$ and $\vec{b}$ are interchanged. Therefore, norms were developed to build standardized, so-called reduced primitive unit cells, where $a \leq b \leq c$ and all angles are either $\leq 90°$ or $\geq 90°$. That means, first it must be ensured that only non-centered cells are considered, centered cells may need to be transformed back into primitive cells, and then you simply sort the lattice vectors according to increasing length and call them a, b, and c. In this way, the cells can now be uniformly and conveniently compared with respect to their metric.

For an exercise, try to specify how the vectors of cell 2 could be obtained using the vectors of cell 3!

We will see, that if we allow to add further lattice points, we get 7 more lattice types giving 14 in total, leading us to the Bravais lattices – as a reminder, we do this, because we want to find the best representative for our crystal system, for the symmetry of the crystal system. However, before being presented in detail, an overview on the different types of centering should be given (◻ Fig. 2.35).

— The primitive cell is the special case that there is *no centering*; this type of lattice is symbolized by the letter **P**.

— Another possibility is that there is one additional lattice point at the center of *one face* of the unit cell, which corresponds to a *single-sided face centering* (which is also called base face centering) and is symbolized by the letter of the face on which the lattice point is placed; according to a convention, this should be the **C** face, that is, the face spanned by the lattice vectors $\vec{a}$ and $\vec{b}$. Alternatively, the face **A** (defined by the

2

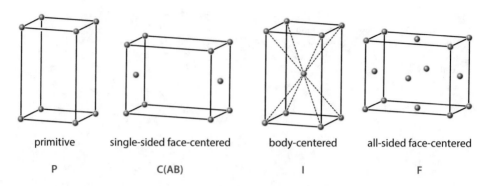

| primitive | single-sided face-centered | body-centered | all-sided face-centered |
| P | C(AB) | I | F |

◘ Fig. 2.35 Overview of the different types of centering

lattice vectors $\vec{b}$ and $\vec{c}$) or the face **B** (defined by the lattice vectors $\vec{a}$ and $\vec{c}$) can also be selected, resulting in an **A** or **B** centering, respectively.

— Furthermore, there is a *body centering* variant, in which an additional lattice point is present exactly at the center of the cell; the symbol for this kind of centering is **I**, derived from "inside."

— Another possibility is that there are additional lattice points at *all faces*, but not inside the unit cell. This is an *all-sided face centering* and the respective symbol is **F**.

— Finally, there is the complicated case of a *rhombohedral centering* (**R**), which is a special case of body centering – we will deal with this kind of this centering separately in ▶ Sect. 2.8.

2.3.2.1 Number of Lattice Points per Unit Cell

From the type of centering, immediately the number of lattice points per unit cell can be deduced: in the primitive lattice type (no centering, **P**), lattice points are only at the corners. Each lattice point, however, belongs to eight unit cells simultaneously, i.e., each lattice point belongs to only one eighth of a unit cell, which results in 8 × 1/8 lattice points = 1 lattice point (◘ Fig. 2.36).

In the case of the one-sided face centering **C**, there are, of course, also eight lattice points at the corners, but additionally two lattice points on one of the faces. These lattice points belong to two unit cells at the same time, i.e., they only belong to one half to each unit cell. The number of lattice points per unit cell is therefore 8 × 1/8 + 2 × 1/2 = 2. A one-sided face-centered cell must therefore also have twice the volume in comparison to a primitive one (◘ Fig. 2.37).

Regarding the body centering **I**, we have eight lattice points at the corners and additionally one lattice point in the center of the unit cell, which belongs only to this single unit cell. Therefore, the number of lattice points is 8 × 1/8 + 1 = 2. A body-centered cell is therefore just as large as a one-sided face-centered.

Finally, in the case of **F** centering, there are eight lattice points at the corners and additionally six lattice points at the faces of the cell, each of which belongs to one half to a unit cell, which results in 8 × 1/8 + 6 × 1/2 = 4 lattice points per unit cell. The **F** centered cell is twice as large as a body-centered cell and four times larger than a primitive cell.

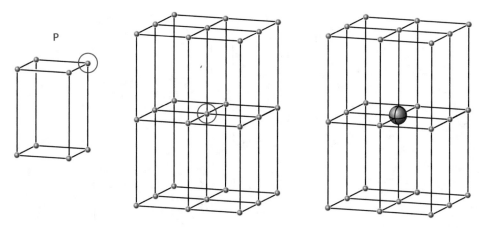

Fig. 2.36 The number of lattice points in a primitive unit cell is one

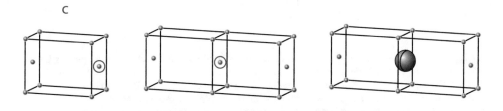

(8 corners x 1/8) + (2 faces x 1/2) = 2 lattice points per unit cell

Fig. 2.37 In the case of single-sided face centering, a unit cell contains two lattice points

2.3.3 **The 14 Bravais Lattices**

As we saw in the previous section, there are four different types of centering (**P, C, I, F**). Of course, centerings have no influence on the symmetry of the lattice. They are only used to better represent the existing symmetry of a lattice of a given crystal system. There are 7 primitive lattices, and if we take into account centerings, there are further 7 centered lattice types, resulting in 14 fundamentally different lattice types in total, the 14 Bravais lattices (also called translation lattices).

The number 14 does not seem obvious at first glance. If we have 7 crystal systems and 4 kinds of centerings, then the total number of different cells including centerings should be 28. In fact, there are only 14 principally different lattice types (in three-dimensional space), as Auguste Bravais, a French physicist, crystallographer, and polymath, was able to show in 1848 [7]. In ◻ Fig. 2.38 all 14 Bravais lattices are shown in a table-like overview, whereby the crystal systems form the columns of this table and the centerings the rows. Obviously, not for all crystal systems all kinds of centering exist – some entries in ◻ Fig. 2.38 are absent. For the triclinic crystal system, for example, there is only the primitive lattice type, for the monoclinic only the primitive lattice and the one-sided face-centered variant exists, while all types of centering are conceivable for the orthorhombic system, etc.

2

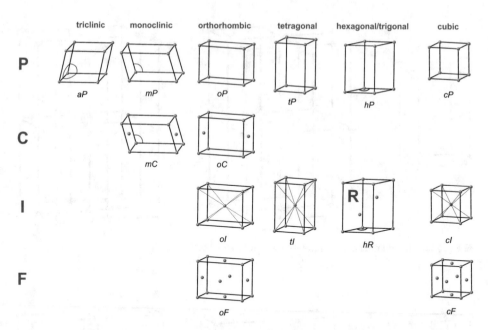

Fig. 2.38 Overview of the 14 Bravais lattices, arranged according to the seven crystal systems. Below the unit cells the corresponding symbols are indicated, which consist of the first letter for the crystal system and a large letter for the type of centering. In order to avoid confusion between triclinic and tetragonal, the old term "anorthic" was chosen for the triclinic system, the corresponding symbol for triclinic-primitive is therefore *aP*. The symbol *hP* is used for both primitive lattices of the trigonal and hexagonal system

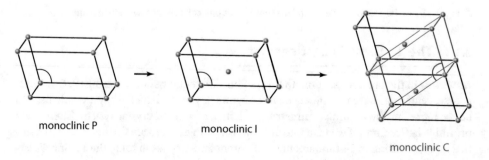

Fig. 2.39 A body-centered monoclinic lattice is possible, but is not considered an independent lattice type, since it can be transferred into a **C**-centered lattice (red cell) by redefining the axes. Note: the (usually oblique) angle *β* may change

There are two reasons why not all types of centering exist for all crystal systems. First, some of the 28 possible lattice types are redundant – meaning that some pairs of lattice types can be mutually interconverted so that this pair does not represent two lattice types, but only one. An example of this redundancy is illustrated in ▪ Fig. 2.39. Starting with the primitive lattice type of the monoclinic crystal system, by adding a lattice point inside the cell, a monoclinic body-centered lattice type can be constructed – an entry that is missing from the overview in ▪ Fig. 2.38. The reason for this is that the

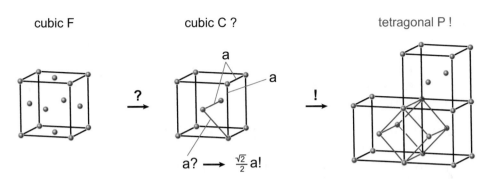

□ Fig. 2.40 For reasons of symmetry, a single-sided face-centered cubic lattice type cannot exist (see text)

body-centered variant can be converted into the monoclinic **C**-centered type. To do this, you only have to redefine two of the axes in such a way that the additional lattice point is now located on the **C**-face, which is spanned by the *a* and *b* axis. Therefore, the **I**-centered lattice type is not required as an independent variant in the monoclinic crystal system. However, it is important to remember that it is not forbidden to choose the body-centered monoclinic unit cell.

The second reason for the fact that not all kinds of centering exist for all crystal systems is due to symmetry. Certain types of centering are forbidden in certain crystal systems for symmetry reasons. Let's take a look at the example in □ Fig. 2.40, which examines the question of whether a **C**-centered cubic lattice type is possible in addition to the **F**-centered cubic lattice type. This is not the case, because a **C**-centering is incompatible with the cubic symmetry. The consideration which lead to this conclusion is as follows: if we remember that there must also be a primitive unit cell for each centered cell, then it should be possible to construct a primitive cell from the **C**-centered cell. This is shown in the right-hand part of □ Fig. 2.40. For reasons of symmetry, however, the three lattice vectors in the cubic crystal system must be identical, and this also means that the distance between equivalent lattice points must be equal. However, as shown in the middle part of □ Fig. 2.40, this is no longer the case for a cubic **C**-centered cell. In fact, a primitive lattice of the tetragonal crystal system is created (shown in red).

Now, the table-like overview from □ Fig. 2.38 is almost completely explained, but the open question remains as to what this somewhat strange **R**-centering is all about. In order to understand the hexagonal crystal system, the **R**-centering – which has not yet been mentioned as a kind of centering – and the associated rhombohedral lattice type, it is useful to deal with the topic of fractional coordinates beforehand. The topic of rhombohedral centering is then discussed further in detail in ▶ Sect. 2.8.

2.4 Atom Positions: Fractional Coordinates

Fractional coordinates can be used to conveniently specify the positions of the atoms in the unit cell. The respective crystallographic axis system of the crystal is used, i.e., the lattice vectors form the basis for this, and the absolute numerical values of the lattice constants are given as unit lengths, i.e., they are set as 1. The respective specific positions of

2

the individual atoms in the unit cell are now specified by coordinate triples x, y, and z, whereby these values are now specified as corresponding fractions of the lattice constants a, b, and c. This is best illustrated by an example: assuming we have an orthorhombic cell with the lattice constants $a = 5$ Å, $b = 20$ Å, and $c = 15$ Å, and an atom of the crystal in this cell would have the absolute coordinates $x = 2.5$ Å, $y = 10$ Å, and $z = 7.5$ Å, then the following fractional coordinates result:

$$x = 2.5 \text{Å} / 5 \text{Å} = 0.5, \ y = 10 \text{Å} / 20 \text{Å} = 0.5, \text{and } z = 7.5 \text{Å} / 15 \text{Å} = 0.5.$$

This means that the relative or fractional coordinates of an atom are obtained by dividing its absolute coordinates by the lengths of the lattice constants. In this example, the atom sits exactly in the center of the unit cell (◘ Fig. 2.41). If a second and a third atom had the absolute coordinates (5 Å, 10 Å, 0 Å) and (2,5 Å, 20 Å, 7,5 Å), the fractional coordinates would be (1, ½, 0) and (½, 1, ½), respectively. Atom 2 would sit on the center of an edge along the b axis, and Atom 3 would sit in the center of the B face (◘ Fig. 2.42 and 2.43). Thus, the advantage of using fractional coordinates over absolute ones should have become clear: we immediately recognize whether an atom is located on so-called special positions, i.e., whether it is located at selected places, such as in the center, at a corner, or at the center of a face of the unit cell. And we recognize this without ever having to look at the lattice constants. This is visualized in ◘ Fig. 2.44 once again in general form: each coordinate triple in which only zeros or ones occur must belong to an atom at the corners; a triple, in which two times ½ occurs, no matter where in the triple, and additionally once 0 or 1, corresponds to an atom that sits in the center of a face and a triple, in which once ½ and additionally only zeros or ones occur, means that the atom is located at the center of an edge. If a coordinate triple does not contain any special values such as 0, ½, 1, etc., then the corresponding atom is located on a so-called general position (x, y, z). The exact definition of special and general positions is derived from whether the atom is located on a symmetry element or not; we will deal with this in more detail later on.

◘ **Fig. 2.41** An atom with the fractional coordinates (0.5, 0.5, 0.5) is located exactly in the center of the unit cell

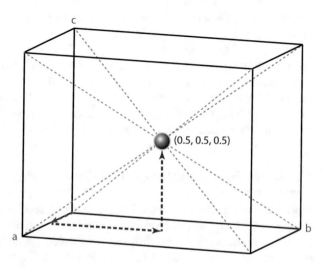

2.4 · Atom Positions: Fractional Coordinates

◻ Fig. 2.42 An atom with the fractional coordinates (1, 0.5, 0) is located at the center of the edge along the *b* axis

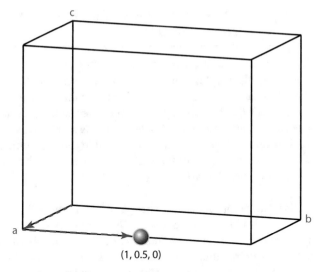

◻ Fig. 2.43 An atom with the fractional coordinates (0.5, 1, 0.5) is located at the center of the B face (which is spanned by the lattice vectors $\vec{a}$ and $\vec{c}$)

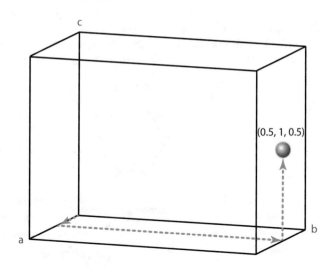

◻ Fig. 2.44 General representation of fractional coordinates with selected atoms on special positions as well as the general position (*x, y, z*)

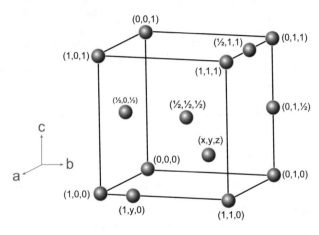

2

2.4.1 Fractional Coordinates Smaller than Zero or Larger than One

Occasionally, a few atoms of a crystal structure may have negative fractional coordinates or those larger than 1. How can that be, when all atoms should be in the (same) unit cell and thus the fractional coordinates should be limited to the range of 0 to 1? Obviously, such atoms do *not lie inside the same* unit cell. The reason for the assignment of such coordinates to certain atoms is that there are two principles according to which the coordinates should be selected:

1. If possible, all atoms should be located in the same unit cell.
2. At the same time, the atoms are to be placed in such a way that meaningful chemical units (molecules) are obtained.

The second point may sound strange at first, because the atoms do have certain places in the unit cell and you cannot choose them completely freely. But, for example, it is allowed to place them at symmetry-equivalent places; and this allows us, for example, to add or subtract 1 to fractional coordinates, either only for x, or for x and y, or even for all three coordinates. In other words, pure translation by any number of whole unit cells is always allowed in all directions (a, b, and c). This means that we are free to place atoms (virtually) in neighboring cells if this is chemically meaningful; an example illustrates this in ◘ Fig. 2.45. In general, however, there is an agreement that at least the majority of all atoms of a crystal structure should be located in the same unit cell.

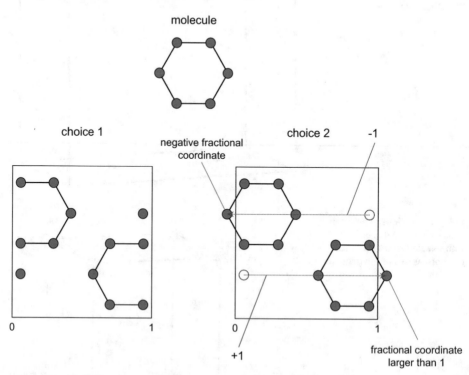

◘ **Fig. 2.45** Example for the different choice of fractional coordinates of a molecule. On the left all atoms are located in the same unit cell. However, it is also possible to place the atoms by subtracting or adding whole units of unit cells so that the chemically meaningful unit (the molecule) appears non-fragmented. This results in fractional coordinates smaller than zero or larger than one

2.5 Exchange of Crystallographic Information (CIF Files)

■ Figure 2.46 shows an impressive chalcanthite crystal formation, chemically speaking copper sulfate pentahydrate ($CuSO_4 \cdot 5\ H_2O$), formerly also known as copper vitriol. In order to fully describe its atomic structure, we need the lattice parameters, the atomic positions, and the so-called space group (▶ Chap. 5); this information is given together with the crystal system in ■ Table 2.1; the hydrogen atoms are omitted for the sake of clarity.

Before 1990 it was common practice to publish and share crystal structures in this table form. However, it would be very tedious if we had to discuss crystal structures exclusively in the form of a table filled with numbers and if we had to imagine the arrangement of atoms in space. Nowadays, crystal structures are exchanged digitally or electronically in the form of so-called CIF files (CIF, crystallographic information file). This is an ASCII text-based file format that contains all relevant information about a crystal structure. In the meantime, there are many programs that can interpret CIF files and visualize the corresponding crystal structure in 3D! This simplifies the analysis and description of the atomic structure considerably.

2.6 Visualization of Crystal Structures with VESTA

A software that is especially suitable for visualizing crystal structures and is also freeware is VESTA [8], which is available for all popular operating systems (Windows, Mac OS, and Linux). This program does not only allow the visualization of existing data or the import of CIF files – it allows the construction of arbitrary crystal structures, provided that one has the information regarding the space group and the fractional coordinates of the atoms of the crystal.

■ **Fig. 2.46** Crystals of copper(II) sulfate pentahydrate. (CC BY-SA 3.0, Wikipedia user ▶ Stephanb)

2

■ Table 2.1 Crystallographic information on copper sulfate pentahydrate (chalcanthite)			
Space group	$P\bar{1}$		
Crystal system	triclinic		
a	5.9553 Å		
b	6.1084 Å		
c	10.7048 Å		
α	77.4090°		
β	82.3720°		
γ	72.6740°		
Atomic coordinates	**x/a**	**y/b**	**z/c**
Cu1	0.5	0.5	0.5
Cu2	0.5	0	0
S1	0.12527	− 0.01315	0.28634
O1	0.64893	− 0.28899	0.11748
O2	0.65182	0.18247	0.07346
O3	1.12844	− 0.43479	0.12435
O4	0.20280	0.46556	0.59358
O5	− 0.11557	− 0.04315	0.30159
O6	0.13680	0.13959	0.37259
O7	0.29749	− 0.24436	0.31756
O8	0.17223	0.09305	0.15153
O9	0.48076	0.75491	0.58387

In this tutorial-like section, we want to get to know the software and see how we can construct unit cells and add atoms to a crystal structure and what possibilities of 3D representation of the resulting structure exist.

2.6.1 Download and Installation

Visit ► http://jp-minerals.org/vesta/en/ and choose the item Download on the left menu (■ Fig. 2.47). On the next page, click on the appropriate package for your operating system, and then the download of a compressed archive file (for Windows: ZIP) will start. Your download folder now contains the archive file, which only needs to be extracted to a folder of your choice. This folder now contains the executable file (under Windows: VESTA. exe); a double-click will start VESTA (■ Fig. 2.48).

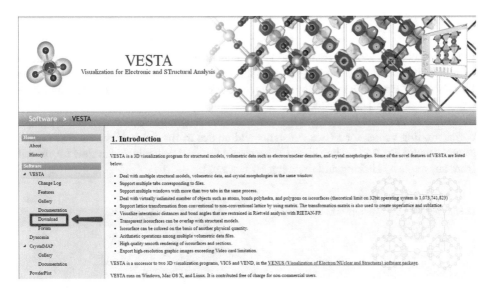

Fig. 2.47 Source of supply of the VESTA archive from the homepage

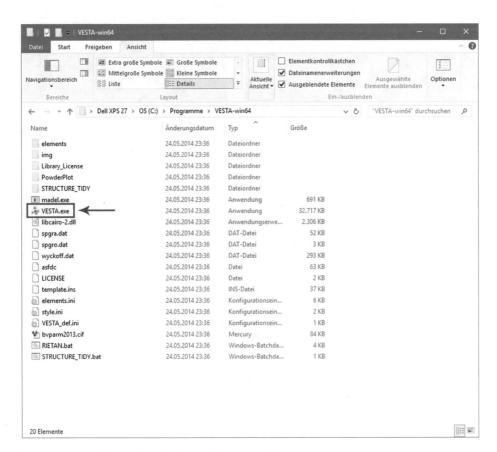

Fig. 2.48 A double-click on the file VESTA.exe starts the program

2

2.6.2 Building the Unit Cell

In this example we want to create the crystal structure of chalcanthite (i.e. copper sulfate pentahydrate) on the basis of the information given in ▪ Table 2.1. Select *File – New Structure…* (▪ Fig. 2.49); in the appearing window you can enter a name in the tab *Phase* (e.g., "Chalcanthite") (▪ Fig. 2.50).

Switch to the *Unit Cell* tab, set P-1 under the *Space group* selection box and enter the corresponding values for lattice constants and angles in the fields of the section *Lattice parameters* (▪ Fig. 2.51). Then click OK to create the unit cell, which can now be viewed from any angle by clicking and dragging (▪ Fig. 2.52).

▪ **Fig. 2.49** A new structure is being created

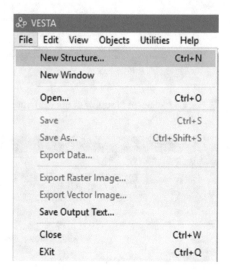

▪ **Fig. 2.50** Under the tab *Phase* the structure can be given a name

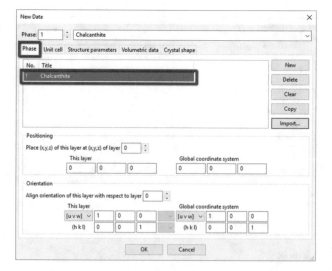

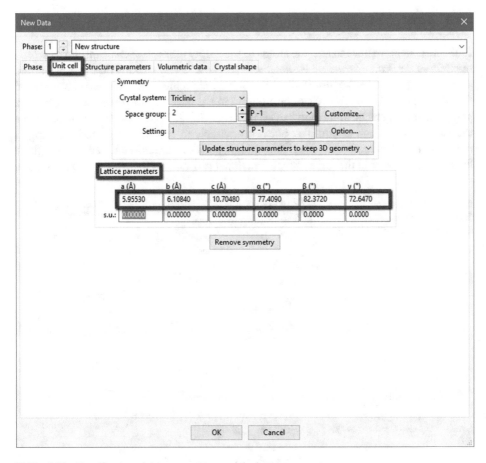

Fig. 2.51   Specification of the space group and the lattice parameters

Fig. 2.52   The empty unit cell of chalcanthite

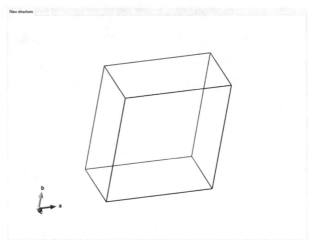

2

2.6.3 Specifying the Atomic Structure

In the next step, we have to add the atoms to the empty unit cell of chalcanthite by selecting *Edit – Edit Data – Structure Parameters...* (▢ Fig. 2.53). By clicking on the *New* button, a new atom (▢ Fig. 2.54) is inserted, which is still unspecified and automatically assigned the coordinates (0,0,0). This has to be adjusted in the next step. If you click on *Symbol*, a periodic table opens in which you can select Cu (▢ Fig. 2.55). In principle, you can name the newly added atom as you like, but it is advisable to choose a clear and meaningful name under *Label*, preferably "Cu1." The atomic positions are now entered into the three fields directly below in the form of their fractional coordinates (0.5, 0.5, 0.5) (▢ Fig. 2.56).

▢ **Fig. 2.53** The command *Structure Parameters...* opens the dialog box for adding atoms

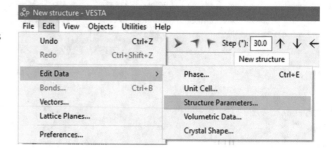

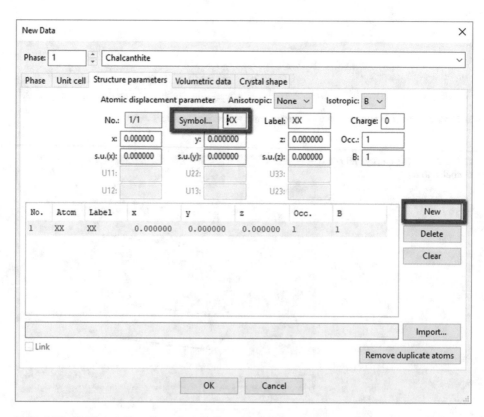

▢ **Fig. 2.54** The button *New* allows you to add atoms. A click on *Symbol...* opens the periodic table

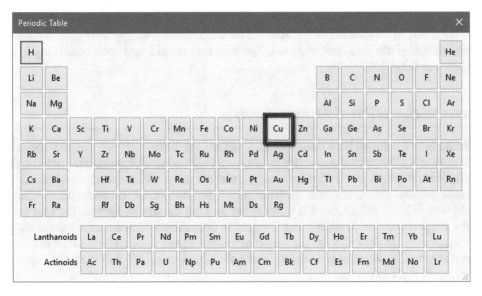

Fig. 2.55 A click on "Cu" selects the atom type for the first atom from the periodic table

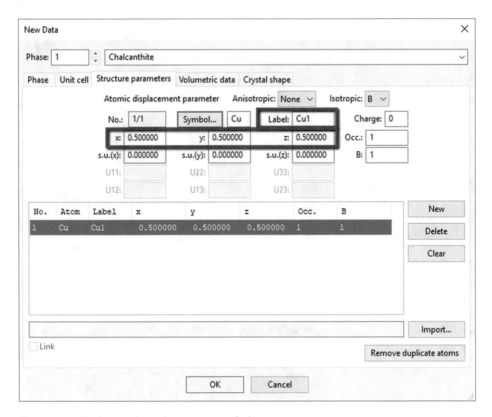

Fig. 2.56 The fractional coordinates are specified

2

A click on *OK* will transfer the entry into the structure. Repeat the last steps for all other atoms. Hydrogen atoms are often not included, because they quickly make the representation of the structure cluttered. The structure should now look something like shown in ◘ Fig. 2.57.

2.6.4 **Generation of Atomic Bonds**

So far only the atoms have been specified, but not the bonds between them. A special feature of VESTA is that these have to be created manually, while other programs usually generate them automatically based on certain rules, which specify at which distance atomic pairs are still considered to be connected to each other. This characteristic of VESTA quickly turns out to be a big advantage, because the user has the greatest possible flexibility and freedom to determine which atoms are connected to each other. In chalcanthite, the copper atoms are bound to six oxygen atoms and the sulfur atoms to four oxygen atoms. Bonds are added using the *Edit – Bonds...* command (◘ Fig. 2.58).

◘ **Fig. 2.57** The unit cell of the chalcanthite filled with all atoms (except the hydrogen atoms)

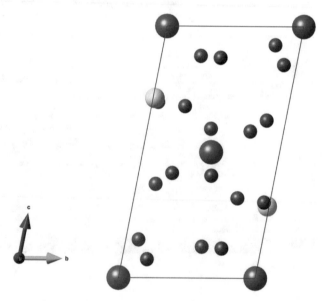

◘ **Fig. 2.58** The command to specify bonds

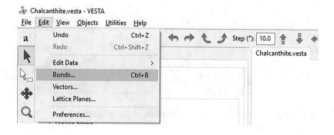

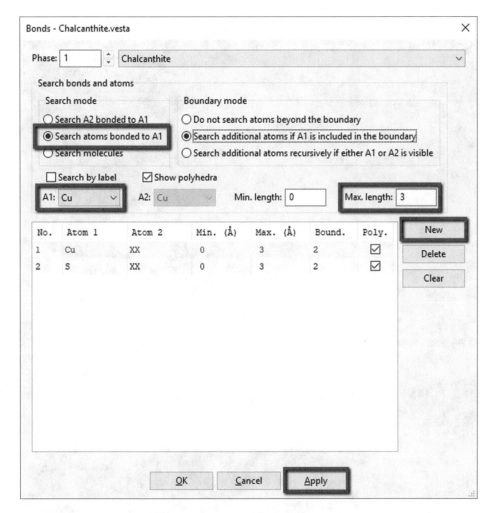

Fig. 2.59 The dialog box for specifying the bonds which originate from an atom

Click *New*, select the option *Search Atoms bonded to A1* and enter 3 for *Max. length*. A click on *Apply* will generate the bonds (Fig. 2.59). Repeat the appropriate steps for the sulfur atom and then close the dialog box by clicking *OK*.

2.6.5 Different Display Styles, Boundaries, and the Export of Images and CIFs

Once the bonds have been created, it is worth trying out the different display styles that can be accessed via the *Style* check box. For the representation of the crystal structure in the form of coordination polyhedra, select *Polyhedral*, whose appearance can be further changed either via the *Objects tab* or by clicking the button *Properties…* (Fig. 2.60).

2

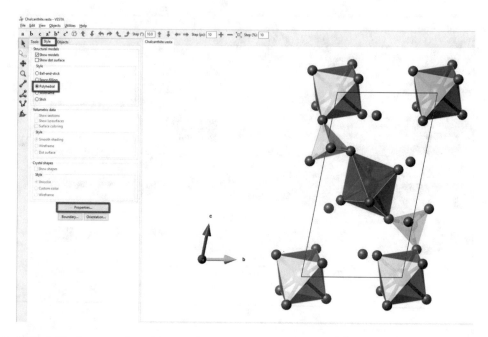

◘ Fig. 2.60   Representation of the crystal structure as coordination polyhedra

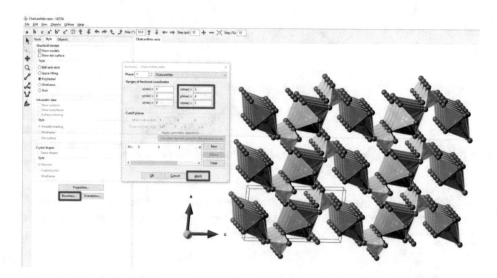

◘ Fig. 2.61   Specification of how many unit cells are to be displayed along the three axes

Familiarize yourself with the program's many options and embark on an atomic journey through your crystals!

It is also possible to use the command *Boundary...* to display not only a single but multiple unit cells; the appearing dialog box can be used to specify in which range the fractional coordinates *x*, *y*, and *z* are to be displayed (◘ Fig. 2.61).

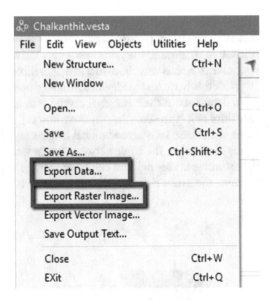

Once you have found a representative representation of your crystal structure, you can generate high-resolution images of it, and you can export the crystal structure according to the CIF format and share and distribute it electronically; both commands can be found in the *File* menu (■ Fig. 2.62). VESTA is also able to depict crystal shapes, i.e., morphologies – see ► Chap. 3 for the respective instructions.

2.7 Four Aspects of the Hexagonal (Crystal System)

In this section we will look at hexagonal patterns and especially at the hexagonal crystal system. The hexagonal crystal system is somewhat more difficult to handle than all the others. Crystallographers sometimes jokingly say it was invented by the devil. Maybe not, but the author would agree that the hexagonal (but also the trigonal) crystal system is more complicated than the others. To examine the hexagonal crystal system, it is advantageous to understand what a symmetry operation is. This is a slight anticipation of ► Chap. 3, in which all symmetry elements and operations are discussed in detail, but at this point it is sufficient to realize what the quintessence of a symmetry operation is at all.

A symmetry operation can be understood as follows: we take an arbitrary object, for example, a cube. Then we close our eyes and someone else performs a geometric operation on this object, for instance, a rotation of 90° around a certain axis. If we then open the eyes again and cannot detect any changes to the object – then this geometric operation was a symmetry operation: a symmetry operation leads to an indistinguishable (note the difference between identical and indistinguishable) state of the object at which this operation was performed.

■ Figure 2.63 shows such a symmetry operation on a cube. A rotation of 90° along the drawn axis results in an identical appearance of the cube. The axis around which it was

2

rotated is the symmetry element of the respective symmetry operation, in this case a four-fold axis of rotation. In contrast, ◻ Fig. 2.64 shows a cube in which the same operation – a rotation of 90° around the same axis – obviously does not lead to an indistinguishable configuration, which is why the associated operation does not represent a symmetry operation.

With this basic knowledge of symmetry operations, we should now be able to discuss a few details about the hexagonal crystal system. ◻ Figure 2.65 shows a hexagonal prism. One third of it represents the primitive unit cell of the hexagonal crystal system, where the edges of the base face have identical lengths ($a = b$), while the height (c) is not restricted to any fixed value; the angle between the two edges of the base face is 120°. If we look at the primitive hexagonal unit cell from above, the projection shown in ◻ Fig. 2.65 results, in which the lattice points of the unit cell are drawn as blue circles.

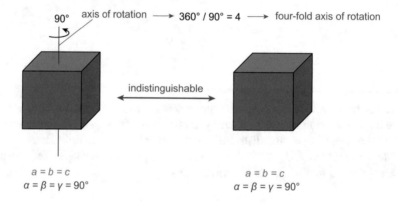

◻ **Fig. 2.63**   A symmetry operation leads to an indistinguishable configuration of an object, here a rotation of 90°

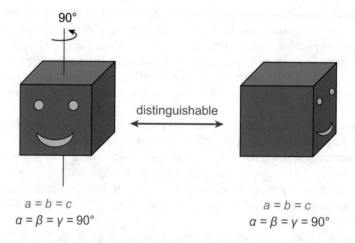

◻ **Fig. 2.64**   Example of an operation that is not a symmetry operation

◘ **Fig. 2.65** Hexagonal prism (left) and its projection onto the plane with four drawn lattice points of a primitive hexagonal unit cell (right)

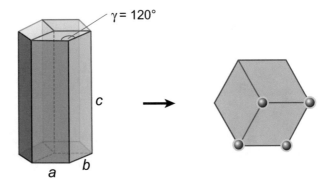

$\gamma = 120°$

c

a   *b*

The following questions should be addressed:

? 1. How are hexagonal unit cells correctly assembled into a larger ensemble?
2. To how many unit cells does a lattice point at a corner belong to?
3. Where in the unit cell is actually the hexagonal crystal system's characteristic six-fold axis of rotation located?
4. Is the honeycomb pattern compatible with a translation lattice?

✔ **Answer to Question 1**

In principle, of course, there is only one way to create a larger crystal made of several unit cells. This consists of assembling unit cells together in all three directions of the coordinate system according to the translation principle (◘ Fig. 2.66). You already know that, of course, and you may wonder why we are discussing this topic again, this time explicitly for the hexagonal crystal system. The reason for this is that there is an obvious difference between the form of the primitive hexagonal unit cell on the one hand and what we generally imagine under the term hexagonal, i.e., something six-sided, on the other hand. How does this fit together?

In ◘ Fig. 2.67a, a hexagonal prism (in the projection) has been drawn into an ensemble of hexagonal unit cells and highlighted in color. Now we can see that this hexagonal prism is not composed of three unit cells; to achieve this would require rotating the unit cells and then joining them together to form a triad (▶ Sect. 1.7). However, this would violate the translation principle – rotations are not allowed!

◘ Figure 2.67b shows that the hexagon rather consists of two complete hexagonal unit cells (light blue shaded) and two further halves of two additional unit cells (dark blue shaded).

✔ **Answer to Question 2**

That may be a strange question for some of you. But drawing hexagons into an ensemble of hexagonal unit cells can be confusing, especially if done in the manner shown in ◘ Fig. 2.68a. One could come up with the idea that the highlighted and with a (1) marked lattice point belongs to three unit cells or maybe six, because we have to take into account that the image is only a two-dimensional projection and we have to consider the third dimension.

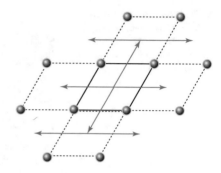

◻ Fig. 2.66   Assembly of primitive hexagonal unit cells according to the translation principle

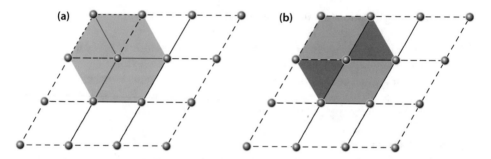

◻ Fig. 2.67   The relationship between a hexagonal prism (**a**) and the primitive hexagonal unit cell (**b**) (see text)

The attentive readership, however, will not be deceived by such tricks: the correct answer to the question is, of course, eight (see ◻ Fig. 2.36). The lattice point belongs to four unit cells in the plane and four more if the third dimension is taken into account (◻ Fig. 2.68b). Note: all unit cells are parallelepipeds. And a lattice point at the corner always simultaneously belongs to eight unit cells, regardless of which concrete crystal system we are looking at.

✅ Answer to Question 3

Intuitively, we are inclined to place the six-fold axis of rotation axis in the center of the hexagonal unit cell. We can verify whether this is correct or not by carrying out a corresponding rotation around an axis passing through the center by 60° and see if a congruent configuration is achieved. ◻ Figure 2.69, which shows a stepwise rotation of the unit cell in steps of 10°, shows that this is obviously not the case!

The correct answer to this question is that the six-fold axis of rotation is located at the corner of the unit cell! As shown in ◻ Fig. 2.70, the position of lattice point (1) (which sits directly on the axis of rotation) does not change at all by a rotation of 60°; lattice point (2) is transformed to lattice point (3), (3) to (4); and lattice point (4), in turn, passes to the lattice point (1'), which already belongs to the adjacent unit cell! You may now think that this is just a trick and does not really represent a six-fold axis of

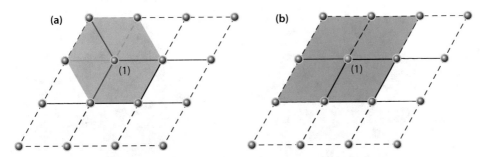

Fig. 2.68 A lattice point at the corner of a primitive hexagonal unit cell does not belong to three (in the plane) or six (3D) unit cells simultaneously (**a**), but to four (in the plane) or eight (3D) unit cells simultaneously (**b**)

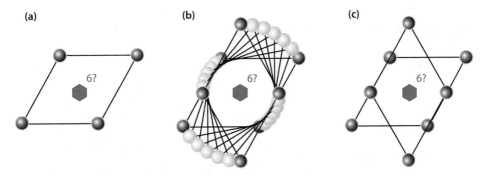

Fig. 2.69 Obviously, a rotation of the primitive hexagonal unit cell by 60° about an axis in the center of the cell does not lead to a congruent final state (**a**), (**c**). In order to be able to follow the rotation more easily, some additional unrealized states are shown (**b**)

Fig. 2.70 The six-fold axis of rotation of the primitive hexagonal unit cell is located directly on the lattice points. In order to be able to follow the rotation more easily, the unrealized state after a rotation of 30° is additionally shown (pale balls)

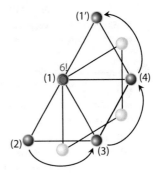

rotation, but that is not the case – the crystal does not care whether the point belongs to the same or a completely equivalent position of an adjacent unit cell, due to the translation principle or because we are dealing with an infinitely extended lattice. One could also say that translations in amounts of whole unit cells have no effect; they are not "detectable."

2

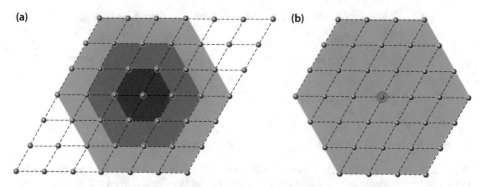

(a)   **(b)**

■ **Fig. 2.71**   The six-fold rotational symmetry of the hexagonal crystal system can be perceived when a larger section of the lattice is displayed (**a**), in particular when the "overhanging" corners are omitted (**b**)

■ **Fig. 2.72**   A section of the honeycomb pattern

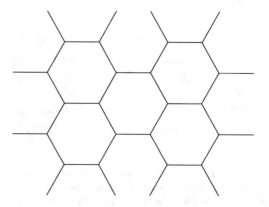

The hexagonal symmetry of the hexagonal crystal system is much easier to perceive when looking at a larger section of the lattice (■ Fig. 2.71a), especially when the "overhanging" corners are omitted (■ Fig. 2.71b).

✔ **Answer to Question 4**

■ Figure 2.72 shows a two-dimensional honeycomb pattern that results when an ensemble of hexagonal prisms is projected along the longitudinal axis. An interesting question is whether or how this obviously periodic pattern can also be described by a translation lattice (Bravais lattice).

The definition of the lattice (▶ Sect. 1.9) says that all lattice points in a lattice must have the same environment. But what exactly does that mean, or which aspects does it include? It refers both to the structural arrangement and – this is important – to the mutual orientation! Well, let's see if these two conditions for the honeycomb pattern are met: the first condition is obviously fulfilled, because no matter which lattice point we consider, it is surrounded by three further lattice points, each arranged at an angle of 120° to the other (■ Fig. 2.73). But the second condition for the presence of a Bravais lattice is not fulfilled! This can be expressed in two ways:

Fig. 2.73   Each lattice point of the honeycomb pattern is surrounded by three further lattice points which are arranged at an angle of 120° to each other

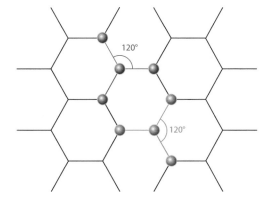

Fig. 2.74   It is not possible to capture all lattice points with two lattice vectors running along the honeycomb walls

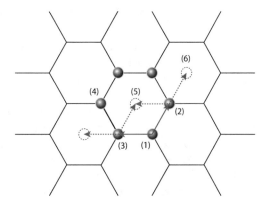

— The pattern looks different depending on which lattice point is selected as the reference point. For example, if we take the central red and green lattice points of ☐ Fig. 2.73 as reference, two different orientations result, which can only be transformed into each other by rotating of 180°.

— If it is a translation lattice, then all lattice points would have to be covered by the corresponding lattice vectors (of a defined direction and length), and for a two-dimensional pattern, we are allowed to use only two lattice vectors. If we start from individual points and perform the translation operation according to these lattice vectors, e.g., from point (1) → (2) and point (1) → (3) (☐ Fig. 2.74), it quickly becomes clear that we cannot reproduce the honeycomb pattern by repeating translations! The lattice vectors applied to point (2) do not lead to point (4), but to the points (5) and (6), which are not included in the original pattern. This means that the honeycomb pattern *cannot* be described with a translation lattice.

The corners of the honeycombs may not be lattice points, but they can be atoms or atomic groups of a crystal – and the corresponding lattice points would sit in the center of the hexagonal honeycombs, forming the hexagonal translation lattice already known to us (☐ Fig. 2.75)! We will address this aspect again in the discussion of graphite (▶ Chap. 7).

2

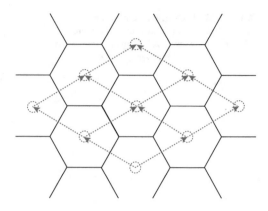

◘ **Fig. 2.75** The centers of the hexagonal prisms of the honeycomb pattern form a primitive hexagonal Bravais lattice

2.8 Trigonal, Rhombohedral, and Hexagonal-R

A special aspect concerning crystal systems and Bravais lattices has not yet been explained: the relationship between the trigonal and hexagonal crystal system on the one hand and the relationship of a rhombohedral lattice to the hexagonal lattice in its R-centering on the other.

The trigonal crystal system is one of the seven crystal systems, and it is characterized by its three-fold rotational symmetry. It is interesting to note that the trigonal crystal system has the same metric as the hexagonal crystal system, although this has a six-fold rotational symmetry as its characteristic feature. Due to the identical metric restrictions, the tri- and hexagonal *crystal system* can be combined to form a common hexagonal *crystal family*, in which identical restrictions concerning the lattice parameters hold: $a = b$, $\alpha = \beta = 90°$, and $\gamma = 120°$ (c remains undetermined).

The confusing aspect of the trigonal crystal system is that there are, in a way, two different kinds of crystal structures: both have in common the maximum three-fold rotational symmetry; however, if you were forced to represent all compounds of the trigonal crystal system with a primitive lattice, you would need two different lattice types, one that looks like assembled of primitive-hexagonal unit cells and another one that has to be described as a rhombohedral lattice with a rhombohedral shape of the unit cell (◘ Fig. 2.76), for which the following metric restrictions apply: $a = b = c$ and $\alpha = \beta = \gamma$.

These relationships have to be clarified, and to do so, the following two questions have to be answered:

— Why can something with three-fold rotational symmetry look like something with a six-fold rotational symmetry?
— Why is there a rhombohedral lattice, but not a rhombohedral crystal system?

To answer question 1, it makes sense to use a two-dimensional projection of the unit cell with hexagonal metric. The reason why something that has only a three-fold rotational symmetry can look as if it has a six-fold rotational symmetry is that the symmetry of the motif is ignored, when it is represented as a lattice point. We should always remember that the lattice point represents a concrete motif, which in most cases has a lower symmetry than the lattice point itself (which always has the maximum symmetry). This can be illus-

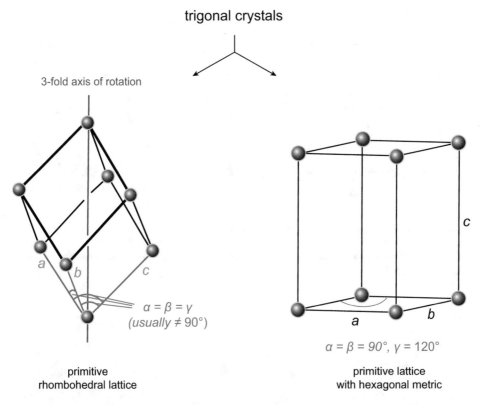

trigonal crystals

3-fold axis of rotation

$\alpha = \beta = \gamma$
(usually $\neq$ 90°)

primitive
rhombohedral lattice

$\alpha = \beta = 90°, \gamma = 120°$

primitive lattice
with hexagonal metric

◻ **Fig. 2.76** There are two types of trigonal crystal structures: those that form a rhombohedral lattice when a primitive unit cell is chosen (left) and those that can be described with a primitive lattice of hexagonal metric (right), although they also have only three-fold rotational symmetry and not a six-fold

trated by a two-stage process: in the first step, we replace the lattice points of a hexagonal lattice with nothing but hexagons – the six-fold rotational symmetry of the lattice is completely preserved (◻ Fig. 2.77a), because the motif itself has a six-fold rotational symmetry (we already know that the axes of rotation pass directly through the lattice points). If, on the other hand, in a second step, we now replace the hexagons with equilateral triangles, motifs that only have a three-fold rotational symmetry, it quickly becomes clear that a rotation of 60° is no longer a symmetry operation. Thus, the associated six-fold axis of rotation has also disappeared – but instead a three-fold axis of rotation is present, and a rotation of 120° represents a symmetry operation (◻ Fig. 2.77b). Nevertheless, the basic arrangement of the triangles in a lattice is compatible with the metric of the hexagonal crystal system. This is simply because a rotation of 120° results in the same configuration as a double rotation of 60° (with six-fold rotational symmetry).

To answer question 2, we need to take a closer look on the rhombohedron as such. A rhombohedron can be derived from a cube, whereby the cube is a special case of a rhombohedron, in which all angles are actually 90°: if we place the cube on one of its tips and now we compress or stretch it along the space diagonal, we obtain a rhombohedron (◻ Fig. 2.78). A rhombohedron is also a parallelepiped, and it is, of course, completely space-filling if you assemble rhombohedra face by face.

2

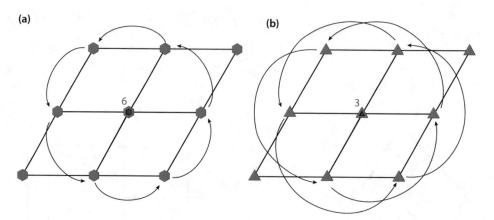

(a)

(b)

□ **Fig. 2.77**   **(a)** Two-dimensional projection of a hexagonal lattice in which the lattice points are replaced by hexagons; the hexagonal rotational symmetry is preserved. **(b)** A hexagonal lattice having only three-fold rotational symmetry because the motifs are now triangles. Here, only a rotation of 120° is a symmetry operation

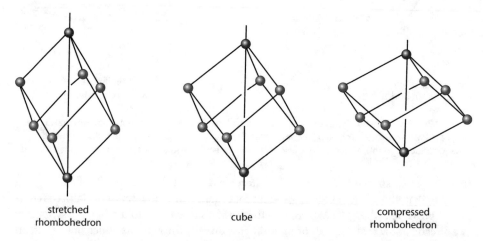

stretched
rhombohedron

cube

compressed
rhombohedron

□ **Fig. 2.78**   Rhombohedron derived by stretching or compressing a cube. The green and blue corner points are transferred into each other by the three-fold axis of rotation

So far, so good – but there is a small problem with this rhombohedron, namely, the fact that none of the three lattice vectors runs along the three-fold axis of rotation (the space diagonal). The problem is that we crystallographers are a little inflexible: crystallographers like strict rules, and one of these rules is that in the tri-, tetra-, and hexagonal crystal system, the axis of rotation of highest order should run parallel to the c axis which is obviously not the case for the rhombohedral lattice! But the rhombohedral lattice can be transformed in such a way that this rule is fulfilled. We'll take a detailed look at that: if the space diagonal is defined as the new axis c', then the lattice points of the unit cell have the fractional coordinates along this c' axis of 0 or 1 for the points drawn in red, 1/3 for the points drawn in green and 2/3 for the points drawn in blue (□ Fig. 2.79a). The green and

(a)   **(b)**

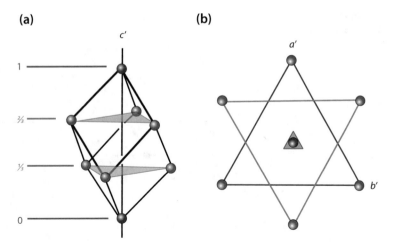

⬛ **Fig. 2.79** (a) Fractional coordinates z for the lattice points of a rhombohedron along the new c' axis; (b) 2D projection of the rhombohedron along the c' axis

blue dots are transferred into each other by the three-fold axis of rotation. In the next step, we consider a projection from above along this c' axis onto the (a',b') plane, the result of which is shown in ⬛ Fig. 2.79b. And now we apply the translation principle along the new a' and b' direction and add further unit cells (⬛ Fig. 2.80).

Now, the question is if it is possible by choosing a larger unit cell, which does not necessarily include only one lattice point, to find one that corresponds to a metric that you already know. Look at the red lattice points! And indeed, the red lattice points define the boundaries of the new unit cell, which according to its metric corresponds to those of the hexagonal crystal system, $a = b$, c is not specified, $\alpha = \beta = 90°$ and $\gamma = 120°$ (⬛ Fig. 2.81). Unlike the primitive hexagonal unit cell, this is a centered cell with two additional lattice points inside the cell, one along c' at 1/3, the other at $c' = 2/3$. This new cell has a volume that is three times larger than the primitive rhombohedral unit cell. In the case of the rhombohedral-centered cell with hexagonal metric (it is also said according to the hexagonal axis system), the axis of rotation of the highest order runs along the lattice vector $\vec{c}$, in perfect accordance with the crystallographic rules! In ⬛ Fig. 2.82, both cells are once again drawn into a single scheme, so that their mutual relation becomes clear: the original, primitive rhombohedral cell and the new, centered cell with hexagonal metric with two additional lattice points inside the cell.

The possibility of achieving a rhombohedral centering shown in ⬛ Fig. 2.82 is the standard option – the so-called obverse setting – but there is a second, fully equivalent one: this is achieved by turning the a and b axis by 60° (parallel to c). This is the *reverse setting*, in which the two additional lattice points inside the cell have the fractional coordinates $(2/3, 1/3, 2/3)$ and $(1/3, 2/3, 1/3)$. ⬛ Fig. 2.83 shows both the obverse and reverse arrangement in comparison.[2] Alternatively, we can also think of the reverse arrangement as being

---

2   The English terms obverse and reverse are originally terms from the coinage system and refer to the front or back of a coin. In a figurative sense, one could say that the obverse and reverse setting of the rhombohedron centering are two sides of the same coin.

2

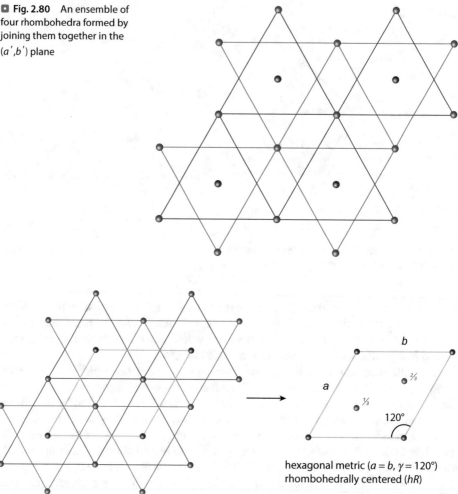

Fig. 2.80 An ensemble of four rhombohedra formed by joining them together in the (a',b') plane

hexagonal metric ($a = b$, $\gamma = 120°$)
rhombohedrally centered (hR)

Fig. 2.81 The red lattice points define the boundaries of the new rhombohedrally centered cell

the result of rotation of the primitive rhombohedron from **Fig. 2.79** by 60, 180, or 300° and then proceeding in a similar way to the procedure in **Fig. 2.80**.

At the end of this section, we will briefly discuss the rotational symmetry of the rhombohedron centering. Since we now have a hexagonal axis system again, one might think that there is also a six-fold rotational symmetry. As **Fig. 2.84** shows, this is not the case. The corner points drawn in red would be mapped on itself, but the green lattice point would come to rest on a blue one and vice versa the blue one on a green one. Because we cannot change the symmetry by choosing a lattice – regardless of which centering we choose; the symmetry is an intrinsic property of the lattice. And this means that – like the primitive rhombohedral cell – the rhombohedral-centered cell with hexagonal metrics can also only have a three-fold rotation symmetry, as **Fig. 2.84** shows.

◻ Fig. 2.82 Comparison of the primitive rhombohedral cell and the new centered cell with hexagonal metric and two additional lattice points inside the cell

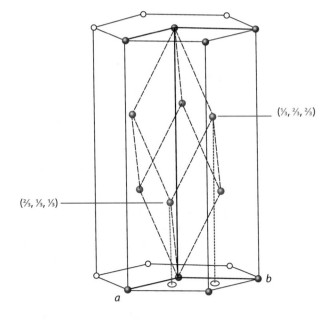

◻ Fig. 2.83 Comparison of the two possible rhombohedral centerings; left: obverse, right: reverse

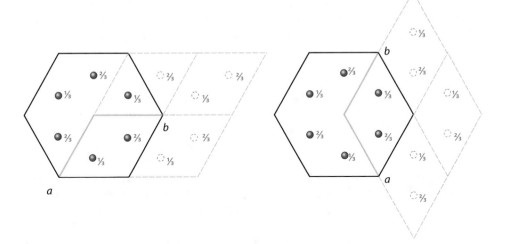

Now we can finally explain why there is a rhombohedral lattice, but no rhombohedral crystal system. A crystal system defines a system that is able to describe all crystals of a certain symmetry. While it is possible – as we have just seen – to transform a rhombohedral system into a one with hexagonal metric, it is not possible to transform the trigonal crystals, whose unit cells in their primitive form look like hexagonal-primitive lattices, into rhombohedral ones. In other words, all trigonal crystals can be uniformly described with cells of the same metric and a maximum order of the rotational symmetry of three:

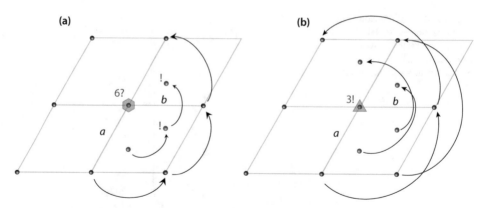

Fig. 2.84 **(a)** The rhombohedral-centered unit cell does not have six-fold rotational symmetry. The corner lattice points drawn in red would be mapped onto each other, but not the green and blue lattice points inside the cell. **(b)** Like the primitive rhombohedral unit cell, the rhombohedral-centered cell has only a three-fold rotational symmetry, so that a rotation of 120° leads to an indistinguishable configuration

always with hexagonal metric; some of them only have to be transformed into the rhombohedral-centered form.

Which substances of the trigonal crystal system are described with the one (hexagonal, rhombohedral-centered, *hR*) or the other variant (primitive-hexagonal, *hP*) are also reflected in the corresponding space group symbols, which are explained in detail in ► Chap. 5.

■ ■ Summary

In the first section of the chapter, the central terms for describing the outer form of crystals were introduced: Tracht ("traditional costume") and habit. Based on this, Miller indices were explained, which allow to clearly identify a set of lattice planes and crystal faces. The outermost lattice planes of a set of lattice planes always form the boundary faces of the crystal.

Subsequently, the reasons for looking at centered variants in addition to the primitive unit cells were discussed. The main advantage is that centered unit cells better reflect the symmetry of the lattice. This led to the derivation of the 14 Bravais lattices.

The electronic exchange of crystal structures takes place by means of CIF files, in which all relevant information about a given crystal structure is stored, including the atomic positions, which are indicated as fractional atomic coordinates. A concrete example showed how crystal structures based on the CIF format can be visualized.

In two subsections, this chapter finally dealt with the particularly complicated crystal systems, the hexagonal and trigonal. The structures of the trigonal crystal system are divided into two subspecies: those that can be described by a rhombohedral-primitive unit cell and those that can be described by a hexagonal-primitive cell. Both can be described together in the same crystal system, because the rhombohedral-primitive lattices can be converted into rhombohedral-centered lattices with hexagonal metric.

References

1. Barnard AS, Russo SP (2009) J Mater Chem 19:3389–3394. https://doi.org/10.1039/b819214f
2. Markov IV (2003) Crystal growth for beginners: fundamentals of nucleation, crystal growth, and epitaxy, 2nd edn. World Scientific, Singapore
3. Sunagawa I (2005) Crystals – growth, morphology and perfection. Cambridge University Press, Cambridge
4. Benz KW, Neumann W (2014) Introduction to crystal growth and characterization. Wiley-VCH, Weinheim
5. Miller WH (1839) A treatise on crystallography. Deighton, Cambridge
6. Marr D (2010) A computational investigation into the human representation and processing of visual information. MIT Press, Cambridge
7. Bravais A (1850) Mémoire sur les systèmes formés par des points distribués regulièrement sur un plan ou dans l'espace. Journal de l'Ecole Polytechnique 19:1–128
8. Momma K, Izumi F (2011) VESTA 3 for three-dimensional visualization of crystal, volumetric and morphology data. J Appl Crystallogr 44:1272–1276. https://doi.org/10.1107/S0021889811038970

Symmetry (Is Everywhere)

© Springer Nature Switzerland AG 2020
F. Hoffmann, *Introduction to Crystallography*, https://doi.org/10.1007/978-3-030-35110-6_3

3

In this chapter, we will immerse ourselves in the wonderful world of symmetry and explore various aspects of the symmetry of macroscopic objects, including, of course, the symmetry of crystals. This means that we will first look at the objects from the outside and disregard their inner structure, which may also include translational symmetry.

We are still in the process of categorizing crystals according to their symmetry in a systematic and hierarchical way, having already managed two steps on the ladder (◻ Fig. 3.1): the 7 crystal systems and the 14 Bravais lattices should already be familiar to us. In this chapter we will reach the next level with an explanation of the 32 crystal classes (also called crystallographic point groups). The last stage will then combine both aspects: on the one hand the macroscopic symmetry properties of crystals and on the other hand their (inner) translational symmetry. This will lead to the 230 so-called space groups, which are nothing more than a complete description of the symmetry of crystalline objects.

The word symmetry comes from Greek and literally means "evenness" or "regularity." In a broader sense, it is also associated with "harmony." And indeed, symmetry is generally perceived as "beautiful." Symmetrical arrangements and patterns are simply stylish!

Before we will look into the various variations of symmetry, let us first point out a formal distinction that is important for the further progress, namely, the distinction between *symmetry operation* and *symmetry element*.

3.1 Symmetry Operation and Symmetry Element

A symmetry operation (SO) is a geometric transformation that maps an object onto itself, something that leads to an indistinguishable configuration, i.e., this operation is carried out on the object, at least imaginary. If we take another look at the example in ▸ Sect. 2.7 (◻ Fig. 2.63), the symmetry operation here consists of a rotation of the cube by 90° around the indicated axis (◻ Fig. 3.2).

◻ **Fig. 3.1** We already climbed two of the four steps on the step ladder to fully systematize crystals. Ready for the next level!

4. level: space groups
complete symmetry

230

3. level: crystal classes
crystallogr. point groups

32

2. level: Bravais lattices
primitive & centered

14

1. level: crystal systems
metric & symmetry of the UC

7

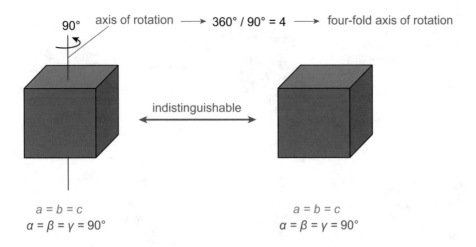

Fig. 3.2 The symmetry operation consists of a rotation around the marked axis by 90°; the symmetry element is the axis, the geometric object at which the operation is performed

The symmetry element (SE) on the other hand is the geometric object on which this operation is carried out – this can be a point, a line, or a plane. That means, in the given example, we rotate the cube by 90° around the axis – and exactly this axis is the geometric object around the rotation is carried out; it is the symmetry element – here a four-fold axis of rotation. This axis comprises all points which do not change their positions while carrying out the symmetry operation – mathematically speaking, it comprises all invariant spatial points. Usually, several different SOs can be carried out on the same SE. For example, we could also rotate the cube from ◘ Fig. 3.2 by 180° around the axis (SO), which would make it a two-fold axis of rotation (SE).

By now you have received an insight what the meaning of point symmetry is: point symmetry means you carry out a symmetry operation, in which at least one point of the object is not moving around. And this is different to translational symmetry, in which translation operations of the whole object are involved.

3.2 The Five Point Symmetry Elements

If we solely consider the external symmetry of macroscopic objects, only five symmetry elements exist. They will be discussed below:
- Identity
- Mirror plane
- Axis of rotation
- Center of inversion
- Rotoinversion axis

3.2.1 Identity

All objects, even the most asymmetric ones, have at least one element of symmetry: the identity. At first, it may seem somewhat contradictory that completely asymmetrical objects also have a symmetry element, but we can also consider the identity as a one-

3

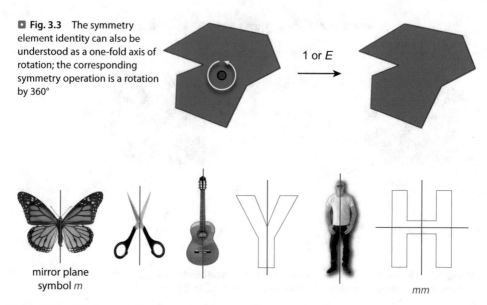

◻ **Fig. 3.3** The symmetry element identity can also be understood as a one-fold axis of rotation; the corresponding symmetry operation is a rotation by 360°

1 or *E*

mirror plane
symbol *m*

mm

◻ **Fig. 3.4**   All these objects show (more or less perfect) mirror symmetry; the right and left half is identical, but mirrored. The letter H has two mirror symmetries simultaneously. The mirror symmetry is often not 100% fulfilled, especially with natural objects. (Monarch butterfly: Didier Descouens, CC BY-SA 3.0, guitar: Martin Möller, CC BY-SA de)

fold axis of rotation. And a rotation by 360° is, of course, a symmetry operation, since the final state of the object is indistinguishable from its original position (◻ Fig. 3.3). In the world of molecules, the identity is identified by the symbol *E*, in the world of crystals simply by a 1, which is an indication of the one-fold rotational symmetry. The element of identity was introduced by mathematicians for formal reasons within the framework of group theory. In this framework it is the neutral element. Apart from the explanation given in ▸ Sect. 6.4 of what is meant by a group, we do not want to deal with group theory here.

3.2.2 **Mirror and Rotational Symmetry**

3.2.2.1 **Mirror Symmetry**

The objects shown in ◻ Fig. 3.4 have something in common. They all have mirror symmetry, which is sometimes also referred to as axis, reflection, or bilateral symmetry. In other words, an object that remains unchanged upon reflection has mirror symmetry. If the object is two-dimensional, the corresponding symmetry element is called mirror line; for three-dimensional objects, it is the mirror plane. Graphically, the mirror plane is symbolized with a simple, continuous line and abbreviated with the letter *m* (from mirror). An object can, of course, also have more than one mirror symmetries, for example, the letter H has a vertical and a horizontal mirror line. The resulting symbol is *mm*. We will discuss more intensively those cases where there are more than one symmetry element is present, whether identical or different, including the question in which order the various symmetry elements are listed.

A particularly interesting aspect of mirror symmetry – outside the crystal world – will be highlighted in the following short excursion.

Symmetry and Beauty of Faces

◪ Figure 3.5 (left) shows the face of Kelly George, a woman who is commonly perceived as beautiful. At least she became the beauty queen of the US state Arkansas in 2007. It is true that beauty is in the eye of the beholder, and there is no question of taste, but there is an interesting connection between symmetry and the beauty or attractiveness of faces. It has been found out that, independent of culture, gender, and age, especially those faces that are particularly symmetrical are perceived as beautiful. How symmetrical a face is can be easily found out by cutting a photo of the face in the middle and then mirroring one half each and glue the two parts together again. This can be done either for the right-hand or left-hand side of the portrait. This results in two exactly mirror-symmetrical Kelly Georges (◪ Fig. 3.5, middle and right). And we see that Kelly George actually has a highly symmetrical face, because (a) the two perfectly mirror-symmetrical portraits do not deviate so much from each other and (b) they do not differ very much from the original picture.

◪ **Fig. 3.5** Kelly George, beauty queen of the US state Arkansas in the year 2007 (left), joined picture by reflection of the left (middle) and right half of the portrait (right). (Friedrich Graf, US Air Force, CC-BY-SA 3.0 de)

But that's not all: if you ask 10,000 people which of the three pictures shows the prettiest face, the majority will reply: the original. So, we remember that a slight(!) deviation from perfect symmetry is perceived as even more beautiful.

This is one of a series of highly exciting insights from anthropologists and sociobiologists who have scientifically addressed the issue of attractiveness and its determining factors in the choice of partners. At this point, the author can especially recommend the work of Prof. Dr. Karl Grammer. An overview of his scientific work can be found at the following URL:

▶ https://univie.academia.edu/KarlGrammer.

Fig. 3.6   An ensemble of gears with various numbers of teeth and different orders of rotational symmetry. (Designed by ► Freepik.com)

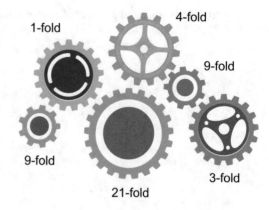

Fig. 3.7   Examples of objects with rotational symmetry. (Three-winged boomerang with kind permission of "Der Linkshänder". Fire lily – Denis Barthel CC BY-SA 3.0. Car rim – courtesy of Scanmore UG. Red wine glass – © mylisa – ► Fotolia.com)

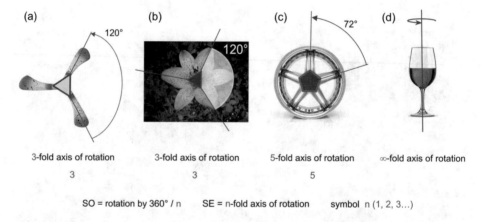

3.2.2.2 Rotational Symmetry

Figure 3.6 shows a series of different cog or gear wheels. These gears have different rotational symmetries of different order between 1 and 21. The number of the teeth, on the other hand, is between 9 and 21 – please note that in order to correctly determine the order of the rotational symmetry not only the number of teeth but also the symmetry of the hubs has to take into account. This picture of cogwheels already shows that ordinary objects can show rotational symmetry of any order.

A three-bladed boomerang (■ Fig. 3.7a) has a three-fold axis of rotation, which means that a rotation by 120° leads to an indistinguishable configuration. A three-fold axis of rotation is graphically represented with a small triangle and simply abbreviated with the number 3 as a symbol. Flowers or, to be precise, the arrangement of petals in flowers is another example where rotational symmetry is very common. The blossom of the flower shown in ■ Fig. 3.7b also shows a three-fold rotational symmetry. However, the symmetry of biological objects is often not quite perfect, not fulfilled to 100% – in contrast to industrial components, which are now manufactured with very high precision. The car rim in ▪ Fig. 3.7c shows a five-fold rotational symmetry, which means that a rotation of 360°/5 = 72° results in an indistinguishable state.

As already indicated, there is no upper limit for the order of rotational symmetry. This becomes immediately clear when we look at the glass of wine in ▪ Fig. 3.7d. A rotation along the longitudinal axis by any angle does not change its appearance, be it 0.0001 or 31.756°. The glass of wine has a rotational axis of infinite order!

3.2.2.3 Combination of Mirror and Rotational Symmetry

Some of the objects shown in ▪ Fig. 3.7 have only rotational symmetry (besides the identity); others have additional mirror symmetry. There are generally much rarer objects, which show only rotational symmetry. Much more common are objects that also show mirror symmetry.

For example, the car rim (▪ Fig. 3.7c) does not only have a five-fold axis of rotation, but also mirror symmetry. The question is: *How many* mirror planes are there? ▪ Figure 3.8 shows five possible mirror planes. But if you look carefully, it turns out that they are all identical! In order to check whether they are identical or not, we can imagine the mirror planes as "cutting planes" that create corresponding fragments. We then examine whether the fragments can be distinguished from those that would be produced by another cutting plane. In this case, we recognize that identical fragments are created regardless of which mirror plane we look at. For this reason, the car rim has only one single, a *unique* mirror plane. Accordingly, the complete point symmetry for the car rim is 5*m* (one five-fold axis of rotation and a mirror plane).

The identification of unique mirror planes shall be examined in another example: a simple, regular hexagon. In addition to the six-fold axis of rotation, there are two (sets of) mirror planes, which are marked as red and yellow lines in ▪ Fig. 3.9. All red mirror planes divide the hexagon into the same two identical pieces and run through two corners of the hexagon. Or in other words, if we pick out one mirror plane, you can generate the other

▪ **Fig. 3.8** This car rim has a five-fold axis of rotation indicated by a pentagon and one unique mirror plane, which is multiplied by the five-fold axis of rotation. (Courtesy of Scanmore UG)

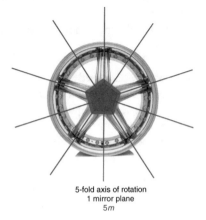

5-fold axis of rotation
1 mirror plane
5*m*

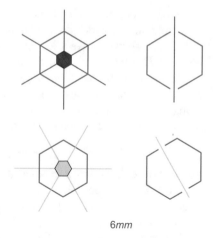

■ **Fig. 3.9**   A regular hexagon has one six-fold axis of rotation and two unique mirror planes: one is running through two opposite corners and the other one through two opposite edge centers

6mm

mirror lines by application of the six-fold symmetry! The rotation of one red mirror plane by 60° or 120° creates the other ones.

❱ Unique symmetry elements are those which cannot be transformed into each other by other symmetry elements.

If we look at the yellow mirror planes of the hexagon, it becomes clear that they, in turn, represent only one single mirror plane of the hexagon, because they can also be converted into each other by the six-fold axis of rotation. However, they form their own set, which is clearly distinguishable from the red one, because the fragments look different: the mirror planes do not run through two corners, but through two opposite centers of edges. But keep in mind that the red and yellow mirror planes are not symmetry-related to each other; they are independent of each other. To create a yellow one from a red mirror plane, you would have to rotate it by 30°. But rotation by 30° is not a symmetry operation of the hexagon!

The full description of the point symmetry of this hexagon is therefore *6mm*.

3.2.2.4 Complete and Short Notation of Symmetry

We have just discussed an example in which the application of certain symmetry elements generates other symmetry elements. Carrying out the six-fold rotation led to the generation of a whole set of equivalent mirror planes from an initial single mirror plane. However, the six-fold axis of rotation did not generate the initially considered mirror plane; it represents a separate symmetry element.

However, there are also cases in which the existence of symmetry elements of an object necessarily leads to the generation of new, independent symmetry elements. Let's look at the letter H (■ Fig. 3.10). The letter H has two vertical mirror planes – but it also has a two-fold axis of rotation axis, which is marked with a green ellipse. This means that the complete point symmetry should actually be *2mm*, but the "2" is sometimes omitted due to a geometric law:

Fig. 3.10 The two perpendicular mirror planes automatically generate a two-fold axis of rotation along their cutting edge, which is indicated by a green ellipse

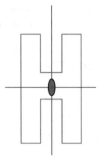

long: *2mm*
short: *mm*

> Each pair of two orthogonal mirror planes generates at their line of intersection automatically a two-fold axis of rotation.

This means that it is not absolutely necessary to specify the two-fold axis of rotation. The complete but redundant point group symbol (long symbol) for the letter H is *2mm*; the so-called short symbol is only *mm*. Especially in the case of crystals, however, there are again some special regulations and exceptions concerning the shortening of long symbols, which will be discussed at the corresponding sections.

3.2.2.5 Rotational Symmetry of Crystals (I)

Above it was stated that macroscopic objects can have rotational symmetry of arbitrary order. In general, this is true – but of all things this is not true for crystals. Interestingly, and in contrast to other macroscopic objects, the rotational symmetry of crystals is limited to the presence of two-, three-, four-, and six-fold axes of rotation, which are shown synoptically in **Fig. 3.11**. There are no axes of rotation of the order five, nor are there any with an order larger than six.

An explanation for this is to be postponed for the time being. We will take up this aspect again in ▶ Chap. 8, in which quasicrystals are discussed. For now, we are registering this as an interesting fact, which is also the reason why the number of crystallographic point groups is limited to 32.

3.2.3 Centrosymmetry (Origin Symmetry, Point Symmetry)

Objects with a center of inversion (also called symmetry center) are constructed in such a way that there are always two corresponding parts, which have the same distance from this symmetry center but are located in exactly opposite directions of space. For example, a regular octahedron has inversion symmetry, because for every point at a corner, there is another corner point, which is just as far from the center but located in opposite directions (**Fig. 3.12**). An octahedron has many other symmetry elements, but there are also objects where the inversion center is the only symmetry element, such as in the playing card

Fig. 3.11   Overview of the axes of rotation in crystallography, with point ensembles of corresponding symmetry, the specification of the rotation angle in degree, and the graphical symbol of the respective symmetry element

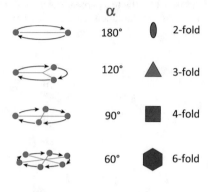

α

180°		2-fold
120°		3-fold
90°		4-fold
60°		6-fold

n = 2, 3, 4, 6

  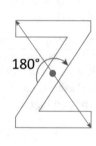 

180°

symbol *i* or $\overline{1}$

Fig. 3.12   Objects with an inversion center. In the case of two-dimensional objects, an inversion is identical with a rotation by 180°

shown in **Fig. 3.12**. All elements of this Jack of Spades card – be it the face, the symbols of spades, etc. – are present in duplicate, and all identical point pairs are symmetry-linked to each other via the inversion center. In mathematical terms, for every point at the coordinates (x, y, z), there must also be an identical point at the coordinates $(−x, −y, −z)$. We can think of a center of inversion also as a point at which you carry out a mirroring, a point reflection. That is why we sometimes speak of point-symmetrical patterns.

As you have probably already noticed, for two-dimensional objects, an inversion center is identical to the presence of a two-fold axis of rotation (rotation by 180°).

The small letter *i* (from Latin *inversio* = the inversion) is used as a symbol for the inversion center in the molecular world. In crystallography it is just a 1 with a line above it, $\overline{1}$. It is pronounced as "one bar," and it is assumed that the original meaning is "minus one" as in other crystallographic notations, too, e.g., Miller indices. However, the origin of this quirky convention, to place the minus sign *above* instead of *in front* of the number or symbol, is an unsolved mystery in the history of crystallography.

3.2.4 **Rotoinversions**

Only one point symmetry element remains to be discussed – the rotoinversion axis (also called rotary inversion axis) at which a rotoinversion is carried out. This shall be illustrated at first on a tetrahedron. A regular tetrahedron is an equilateral three-sided pyramid with an equilateral triangle as base. A tetrahedron can be derived from a cube at which only every second corner is occupied (◘ Fig. 3.13). It is a highly symmetric object that contains two-fold as well three-fold axes of rotation, but unlike an octahedron, it does not have an inversion center. However, this tetrahedron has a four-fold rotoinversion axis. A rotary inversion axis is a symmetry element in which two operations are carried out subsequently; it is a combined symmetry operation: first, a rotation by 360°/n, where n is the order of the rotoinversion axis, and then a point mirroring at the center (an inversion). For the tetrahedron, this coupled operation is shown in ◘ Fig. 3.14: first a four-fold rotation (rotation by 90°) is carried out followed by a point mirroring. The corresponding symbol is $\overline{4}$ ("4 bar"). Graphically it is marked with a filled ellipse enclosed by a hollow rhombus. It is important to note that a tetrahedron has neither an inversion center nor a four-fold axis of rotation. Only the application of the coupled symmetry operation of rotation and inversion leads to an indistinguishable configuration.

Let's now look at further rotoinversions. Crystallographically relevant are only rotoinversions of the order 1, 2, 3, 4, and 6. In ◘ Fig. 3.15 point assemblies are shown, which possess rotoinversion axes of the order 1 to 3. The first two are not particularly interesting, since they can be described as an inversion center ($\overline{1} = i$) or as a simple reflection ($\overline{2} = m$); they do not constitute symmetry elements on their own. However, this is the case with the three-fold rotoinversion axis $\overline{3}$ (◘ Fig. 3.15c). Together with the relationships out of ◘ Fig. 3.14, a further rule can be derived:

> ❯ Rotoinversions with odd order imply the simultaneous presence of a (simple) inversion center.

The already treated and crystallographically relevant rotoinversion of the order 4 is illustrated once again in ◘ Fig. 3.16 together with the remaining six-fold rotoinversion.

Two further laws can be derived from the entirety of the rotoinversions:

◘ **Fig. 3.13**   Derivation of a tetrahedron from a cube in which only every second corner is occupied

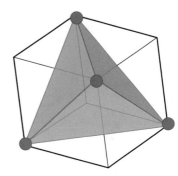

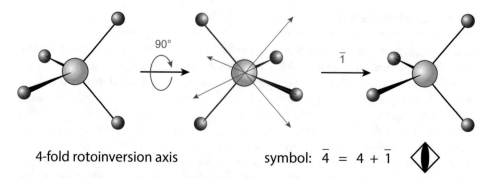

4-fold rotoinversion axis   symbol: $\overline{4} = 4 + \overline{1}$

◘ Fig. 3.14   Four-fold rotoinversion executed on a tetrahedron, which represents a coupled symmetry operation consisting of a rotation by 90° and a subsequent point reflection

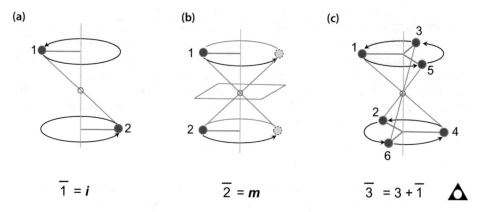

◘ Fig. 3.15   Point patterns (red spheres) showing a rotoinversion axis of the order 1, 2, and 3

◘ Fig. 3.16 The rotoinversions of orders 4 and 6, illustrated by point patterns. A six-fold rotoinversion axis automatically contains a mirror plane perpendicular to the rotoinversion axis

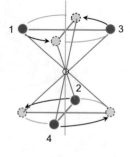

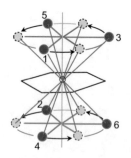

$\overline{4} = 4 + \overline{1}$ $\overline{6} = 6 + \overline{1} = 3 \perp m$

> 1. Even rotoinversions automatically contain an axis of rotation of half of the order: the rotoinversion $\bar{4}$ contains a two-fold axis of rotation, and the rotoinversion $\bar{6}$ always contains automatically a three-fold axis of rotation.

2. If the order of the rotoinversion axis is even and not divisible by 4, then there is automatically a mirror plane perpendicular to this axis – as we only consider crystallographic rotoinversions, this principle holds for the two- and six-fold rotoinversion axis.

In principle, up to this point we dealt with all point symmetry elements. However, it makes sense to discuss another symmetry element that is not used in crystallography, but in chemistry to describe the symmetry of molecules: rotary reflections.

3.2.4.1 Equivalence of Rotoinversions and Rotary Reflections (Hermann-Mauguin vs. Schoenflies)

Concerning symmetry there are in a way two separate worlds: the world of molecules and the world of crystals. Interestingly, during the evolution of natural science, these two worlds were dominated by different scientists, which introduced slightly different approaches to symmetry and systems of nomenclature – which we have to live with today.

On the one hand, there was the German mathematician Arthur Schoenflies, who in 1891 derived and compiled the 230 space groups [1]. Almost simultaneously, the Russian mathematician, crystallographer, and mineralogist Evgraf Fedorov catalogued the 230 space groups [2], whereby Schoenflies and Fedorov stood in lively exchange with each other, but both proceeded methodically largely independently of each other.

The nomenclature of symmetry elements according to Schoenflies was (and still is) used to describe the symmetry of molecules, i.e., finite objects, which show only point symmetry. The nomenclature for describing the symmetry of solids was once again modified by two scientists in the twentieth century: by the German Carl Hermann[1] and the Frenchman Charles-Victor Mauguin.[2] Since then, this symbolism has been used worldwide as a standard in crystallography, i.e., for the description of objects that have not only point symmetry but translational symmetry as well.

Four of the five point symmetry elements are treated identically in both systems, although different symbols are used (◻ Table 3.1), but the fifth one, these rotoinversions just introduced, are expressed in the world of molecules differently, namely, by or as rotary reflections.

It will turn out that both symmetry elements are completely equivalent: each rotoinversion can also be expressed as a rotary reflection. Let's take another look at the example of the tetrahedron. As shown above, this has a four-fold rotoinversion axis, at which first a rotation by 90° and then a point reflection is carried out (◻ Fig. 3.17, upper row). The same symmetry relationship can also be expressed as rotary reflection; the corresponding symmetry element is the rotary reflection axis. Here, too, a coupled symmetry operation is carried out, and the first step is identical – a rotation by 90° – but then a reflection on a horizontal mirror plane (symbol: σ_h) is carried out (◻ Fig. 3.17, bottom row). A horizontal mirror plane is one that is oriented perpendicular to the rotation axis

---

1   Physicist and Professor of Crystallography, ∗ 1898 Lehe (Bremerhaven), † 1961 Marburg

2   Professor of Mineralogy, co-founder of the International Union of Crystallography (IUCr), ∗ 1878 Provins, France, † 1958 Villejuif, France

3

◻ Table 3.1 Comparison of the five point symmetry elements according to Schoenflies and Hermann-Mauguin

Schoenflies	Hermann-Mauguin (international system)
Identity E	Identity 1
Mirror plane σ	Mirror plane m
Axis of rotation C_n (n = 1, 2, 3,...∞)	Axis of rotation n (n = 1, 2, 3, 4, 6)
Center of inversion i	Center of inversion $\bar{1}$
Rotary reflection axis S_n	Rotoinversion axis $\bar{n}$ (n = 1, 2, 3, 4, 6)

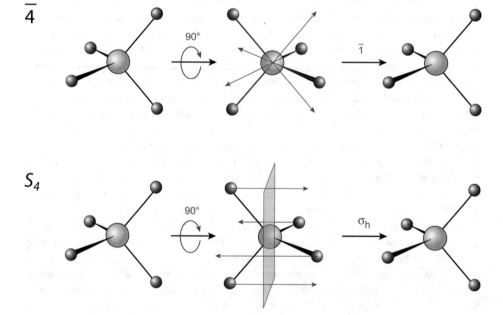

◻ **Fig. 3.17** Four-fold rotoinversion and four-fold rotary reflection in comparison

or, in other words, whose normal vector points in the same direction as the rotation axis (the axis passes "through" the plane at an angle of 90°).

In the case of the tetrahedron, both the four-fold rotoinversion and the four-fold rotary reflection result in identical configurations, which are indistinguishable from the initial state. The order is identical in both cases. However, this is not always the case. ◻ Figure 3.18 shows three arrangements of locomotives which possess rotoinversion or rotary reflection symmetry, respectively. The question is what the order of the respective symmetry element is.

In the first example (◻ Fig. 3.18a), a three-fold rotoinversion axis is present. This means that we pick up a locomotive, first turn it around by 120° and then perform a point reflec-

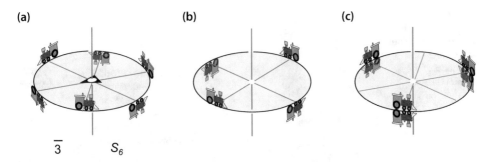

(a) **(b)** **(c)**

$\overline{3}$ S_6

☐ **Fig. 3.18** Three arrangements of locomotives with different rotary inversions and rotary reflections. The arrangement in (**a**) can be described either with a three-fold rotoinversion axis or a six-fold rotary reflection axis. Try to identify the order of the axes for the arrangements in (**b**) and (**c**)!

tion at the center – this obviously leads to an indistinguishable location. We can do this with any locomotive of the arrangement. If we now want to express this with the help of a rotary reflection axis, we have to take a six-fold rotary reflection axis (S_6), which implies a rotation of 60° and then mirroring this locomotive at a plane perpendicular to this axis.

❓ Try to determine the order of (a) the rotoinversion and (b) the rotary reflection of the two arrangements of locomotives shown in ☐ Fig. 3.18b, c. The solution can be found in Appendix A.1.1.

3.3 The Symmetry of Outer Crystal Shapes: Crystal Classes

In the previous section, we got to know all the point symmetry elements (and operations) and are now able to classify crystals with respect of their outer shape according to their point symmetry; this is a further step up on our stepladder (☐ Fig. 3.19).

The interesting thing is that crystals can sometimes look rather peculiar and have a unique appearance, but from the symmetry point of view, they always belong to 1 of the 32 crystallographic point groups (crystal classes). This means that all crystals in the world can be classified into a maximum of 32 principally different crystal classes; only 32 different outer forms exist. The deeper reason for this lies in the limitation of the rotational symmetry. We already know, for example, that a five-fold rotational symmetry is not compatible with a translation lattice.

There is a simple recipe for classification according to crystal classes:

1. We investigate the outer shape of a crystal and determine all symmetry elements that can be found on this specific crystal.
2. We write down all symmetry elements which are clearly distinguishable from each other, i.e., we restrict ourselves to those which represent an independent symmetry element and are not transformed into each other by other symmetry elements.
3. All crystals that have the same symmetry elements belong to the same crystal class.

Later, we will see that these crystal classes can be further categorized according to their crystal systems.

3

Fig. 3.19 We are ready to climb up another step on our stepladder - the conquest of the crystal classes

4. level: space groups
complete symmetry
230

3. level: crystal classes
crystallogr. point groups
32

2. level: Bravais lattices
primitive & centered
14

1. level: crystal systems
metric & symmetry of the UC
7

3.3.1 First Examples

Let's try to apply this recipe to the following two examples. ◻ Figure 3.20 shows a schematically drawn crystal with a specific outer shape. Which symmetry elements are you able to recognize?

Obviously, there is neither a mirror plane nor an axis of rotation. But there is a center of inversion. This is the only symmetry element here (◻ Fig. 3.21). An inversion center is denoted in crystallography as $\bar{1}$. And $\bar{1}$ is also the symbol of the respective crystal class, whenever there is only a center of inversion present. Each crystal class also has a name, which is derived from the description of the most representative geometric shape of the crystals in this class. The crystal class $\bar{1}$ is therefore also called pinacoidal.[3] Since the crystal class $\bar{1}$ belongs to the triclinic crystal system, the full name is triclinic-pinacoidal.

In nature, this is a relatively rare crystal class, but some of the following representatives belong to the triclinic-pinacoidal class: feldspars with the general composition $(Ba,Ca,Na,K,NH_4)(Al,B,Si)_4O_8$ (the components/elements separated by comma in the brackets can replace each other), which are considered to be the most important rock-forming minerals of the earth's crust, turquoise (the eponym for the color of the same name) with the chemical formula $CuAl_6(PO_4)_4(OH)_8 \cdot 4\,H_2O$, and wollastonite, a calcium silicate of the composition $Ca_3(Si_3O_9)$.

Let's investigate a second example! In ◻ Fig. 3.22, you see another schematic drawing of a crystal shape. Are you able to identify all symmetry elements of this shape?

---

3   gr. pinacoid = board, plank; denotes a form consisting of two parallel surfaces connected by a center of inversion. Since a single pair of faces does not form a closed shape, it can only occur in combination with other faces, e.g., further pinacoids or prisms.

Fig. 3.20 Examine this crystal shape for symmetry elements. Which ones can you recognize?

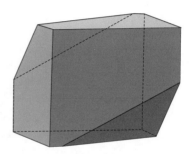

Fig. 3.21 The only symmetry element of this blue crystal shape is the center of inversion

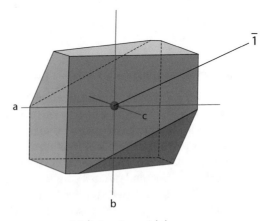

triclinic-pinacoidal

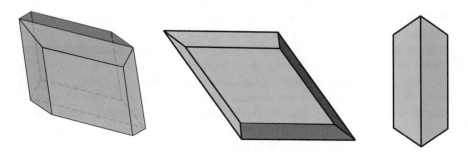

Fig. 3.22 Three views of a crystal shape: perspective view with transparencies (left), side view (center), and front view (right). Which crystal class does it belong to? You can find the answer by identifying its symmetry elements!

There is a mirror plane (m) that divides the green crystal into two identical halves, and there is a two-fold axis of rotation (2) that runs perpendicular to the mirror plane. Figure 3.23 shows two different views of the crystal shape together with the symmetry elements. Under the following URL, you can also find a freely rotating 3D model of this crystal form ► http://webmineral.com/data/Gypsum.shtml.

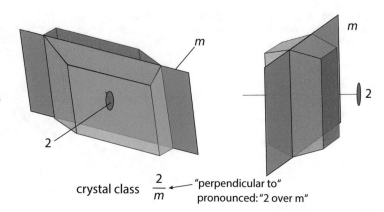

■ **Fig. 3.23** This green crystal contains both a mirror plane and a two-fold axis of rotation as symmetry elements

crystal class $\dfrac{2}{m}$ ← "perpendicular to" pronounced: "2 over m"

The crystallographic notation is $\dfrac{2}{m}$ (pronounced as "2-over-m"). Generally, for one specific viewing direction, rotational symmetry is denoted first, and then mirror symmetry follows. And the horizontal line means that the axis of rotation is running perpendicular (not parallel) to the mirror plane. Or to put it more mathematically, the normal vector of the mirror plane and the axis of rotation are parallel; they have the *same* direction. And because they have the same direction, the two symmetry elements are written one underneath the other and not in a consecutive manner. It's like putting them in the same "column": all symmetry elements appearing in the same direction are denoted into the same column. If, for technical reasons, it is necessary to place the symbol of this crystal class in a continuous text, it may also be written as $2/m$.

Can additional symmetry elements be found at the green crystal? In fact, this is the case: it contains an inversion center (■ Fig. 3.24). But this symmetry center does not need to be mentioned, because it is automatically generated by the other two symmetry elements! Such symmetry elements are not considered in the notation.

The green crystal belongs to the crystal class $2/m$, and the geometric name is prismatic. Since the crystal class $2/m$ belongs to the monoclinic crystal system, the complete name is monoclinic-prismatic. Gypsum (calcium sulfate dihydrate, $CaSO_4 \cdot 2\ H_2O$) belongs to this crystal class $2/m$, for example. ■ Figure 3.25 shows a very beautiful, extremely clear specimen of a gypsum crystal with this typical crystal shape. Other examples of minerals of crystal class $2/m$ are azurite[4] (a basic copper carbonate, $Cu_3(CO_3)_2(OH)_2$), malachite (also a basic copper carbonate, $Cu_2CO_3(OH)_2$, which is often used as a jewellery stone), kämmererite (a chromium-containing clinochlore variety of the composition $Mg_5(Al,Cr)_2Si_3O_{10}(OH)_8$, coveted among mineral collectors), epidote, talc (a very soft, highly water-repellent, greasy feeling mineral with the composition $Mg_3Si_4O_{10}(OH)_2$, which belongs to the group of layer silicates), and vivianite (also known as blue iron ore, which is chemically speaking iron (II) phosphate octahydrate, $Fe_3(PO_4)_2 \cdot 8\ H_2O$, ■ Fig. 3.26).

---

4   Azurite is named after the azure ('sky blue') color of the corresponding pigment. Because of the color of their jerseys, the players of the Italian national football team are also called azzurri - the sky blue.

■ **Fig. 3.24**　This crystal shape also has a center of inversion, but this is not explicitly specified

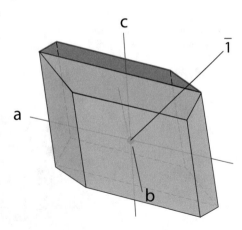

■ **Fig. 3.25**　Transparent gypsum crystal, source: Willow Creek, Nanton, Alberta, Canada. (Rob Lavinsky, ► iRocks.com – CC BY-SA 3.0)

3.3.2　Systematic Approach: Crystallographic Viewing Directions

In the previous examples, symmetry elements occurred in a maximum of one direction. Now let us look at more complicated cases where symmetry elements may be present in more than one direction. This means that the recipe for classifying outer crystal shapes in terms of their symmetry is still the same, but it must be taken into account that the investigation as to which symmetry elements are present follows a systematic approach: we look at the given crystal shape along specific directions (viewing directions). These directions and also their sequence follow fixed crystallographic rules (a systematic overview can be found in ■ Table 3.2). In simple cases, we simply look just along the crystallographic axes *a*, *b*, and *c* and then record for each of the directions which symmetry elements run parallel (for axes) or perpendicular (for planes) to these directions.

The following example will illustrate this: Which symmetry elements can be found in the crystal shape shown in ■ Fig. 3.27? The shape slightly reminds me of a handbag. Apart from the fact that crystals rarely have carrying handles, many handbags may belong to the same crystallographic point group as this crystal.

3

◼ **Fig. 3.26** Green vivianite from the Tomokoni Mine, Potosí, Bolivia. (Size: 4.6 × 4.5 × 3.1 cm) (Rob Lavinsky, ▶ iRocks.com – CC BY-SA 3.0)

◼ **Table 3.2** Overview of the specified viewing directions for the seven crystal systems

	Viewing directions		
Triclinic	*None*		
Monoclinic	*b (c)*[a]	–	–
Orthorhombic	*a*	*b*	*c*
Tetragonal	*c*	*a*	[110]
Trigonal	*c*	*a*	–
Hexagonal	*c*	*a*	[210]
Cubic	*c*	[111]	[110]

[a]In the monoclinic crystal system, symmetry elements can only appear in one direction; by convention this should be the *b* direction; however, in some countries, especially in Eastern Europe, the *c* direction is used instead

Along the *a* direction, we come along a mirror plane *m*, the same is true for the *b* direction, and finally we find a two-fold axis of rotation along the *c* direction (◼ Fig. 3.28). This results in the crystal class *mm*2, which is also called rhombic-pyramidal or orthorhombic-pyramidal according to its basic form; an indication that this crystal class belongs to the orthorhombic crystal system.

◻ Fig. 3.27 Look systematically along the three axes *a*, *b*, and *c*: Which symmetry elements do you discover?

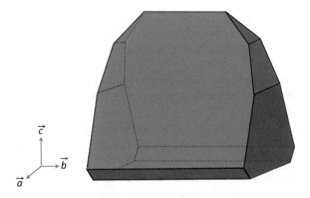

◻ Fig. 3.28 The three symmetry elements that can be found in the purple crystal shape

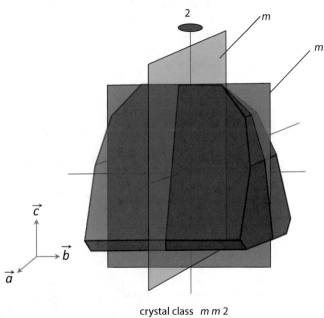

crystal class *m m* 2

viewing direction a b c

However, there are two questions:

1. The first one is devoted to the relative orientation of the crystallographic axis system with regard to the two-fold axis of rotation. The pyramidal side of the crystal could also be aligned parallel to the *a* axis and then the crystal class would be 2*mm*. So, what is correct then? In order to avoid such ambiguities, there is another crystallographic rule: the axis of rotation with the highest order – in this case it is the two-fold axis of rotation – should run along the *c* direction. This defines the crystal class as *mm*2.

2. Is it not sufficient to specify the crystal class as *mm*? The two-fold rotation axis is automatically generated by the two perpendicular oriented mirror planes. In this respect, the explicit specification of the two-fold axis would be redundant. The problem is that there are further symmetry relationships in the orthorhombic crystal system: both 2*m* and *m*2, that is, the combination of a two-fold axis of rotation and a vertical mirror plane in a second viewing direction, automatically generate a further mirror plane in the third direction (try to comprehend this!). This means that 2*m* and *m*2 result equally to *mm*2.

3

Furthermore, there is also the risk of confusion between $2m$ and $2/m$. In order to avoid such possible confusion, the IUCr has decided that in the orthorhombic system, always three symbols or all three viewing directions have to be specified.

Examples of minerals belonging to the crystal class $mm2$ are struvite (◨ Fig. 3.29), a water-containing ammonium magnesium phosphate of the composition $NH_4MgPO_4 \cdot 6\ H_2O$, which has significance in human health because approx. 11% of kidney stones in humans are actually "struvite stones" (in children they account for more than 90% of all kidney stones), and hemimorphite (a water-containing zinc silicate of the composition $Zn_4Si_2O_7(OH)_2 \cdot H_2O$), which means "semi-surface" or "half of the crystal shape." This is due to the fact that the side opposite the pyramidal side of the crystal shape looks different (◨ Fig. 3.30).

◨ **Fig. 3.29** Struvite crystals from a slurry processing plant, image width 7.5 mm. (Walter Kölle – CC BY-SA 3.0)

◨ **Fig. 3.30** Two hemimorphite crystals from a mine in Mexico, in which the "semi-surface shape" can particularly well be seen (size of the crystals approx. 1 mm). (Didier Descouens CC BY-SA 3.0)

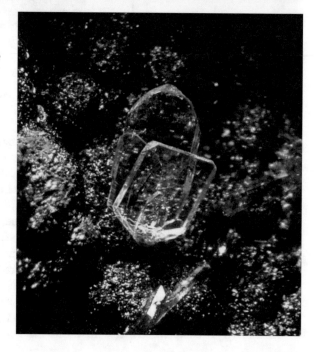

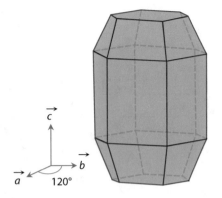

☐ **Fig. 3.31** As in the previous examples, the symmetry elements present should be determined in order to classify this crystal shape according to its crystal class

Crystallographically speaking, the term hemimorphism refers to the removal of a mirror plane perpendicular to the axis of rotation of highest order with respect to the symmetry of the so-called full form (holoedry) of the corresponding crystal system, which for the orthorhombic crystal system would be the point group mmm or, in long form $\dfrac{2}{m}\dfrac{2}{m}\dfrac{2}{m}$.

Let's take a closer look at a last example concerning the classification of outer crystal shapes. ☐ Figure 3.31 shows a crystal shape reminiscent of an old wine barrel. A hexagonal wine barrel that actually belongs to the hexagonal crystal system. In the hexagonal crystal system, the viewing directions are defined as follows: first along the c axis, then along the a axis, and finally along a direction that is not identical to one of the three crystallographic axes, namely, the [210] direction. What does this mean?

3.3.2.1 Directions in the Crystal Lattice

First of all, it is important to note that the directions in the crystal lattice specified in square brackets are not to be confused with Miller indices, which are placed in round brackets. How are the directions derived? To do this, let's first look at ☐ Fig. 3.32, which shows a hexagonal 2D lattice with the lattice vector $\vec{a}$ and $\vec{b}$. A direction in the lattice can now always be achieved by specifying two points, namely, the start and end point. The starting point should always be the origin, i.e., the point $x,y = 0,0$. If the direction vector should not pass through the origin, we achieve this by a translation operation that is always allowed. And now we pick out a certain direction, a vector, and determine the coordinates of that lattice point that the vector hits for the first time along its path. In the given example, this is the lattice point with the coordinates $x,y = 2,1$. The third coordinate is zero, because we are only looking at a 2D example. This results in the coordinate triple $x,y,z = 2,1,0$, which is placed in square brackets: [210].

In the hexagonal lattice there are five other directions that are symmetry-related to the specific direction [210]; they result from rotation by 60° (six-fold rotational symmetry!). The totality of all crystallographically equivalent directions is placed in angle brackets, i.e., the following applies: $<210> = [210]+[120]+[\overline{1}10]+[\overline{2}\,\overline{1}0]+[\overline{1}\overline{2}0]+[1\overline{1}0]$.

Now it should also be clear that the directions along the crystallographic axes a, b, and c can also be specified as [100], [010], and [001].

But now back to our hexagonal wine barrel (☐ Fig. 3.31)! Which symmetry elements can be identified along the three directions $c = [001]$, $a = [100]$, and [210]? Along the c direction there is, of course, the six-fold axis of rotation, and perpendicu-

3

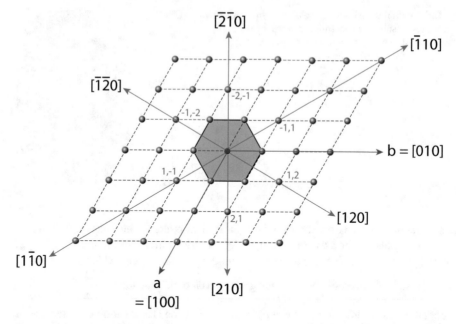

Fig. 3.32 To illustrate directions in the lattice: the directions are specified by corresponding vectors, which always start at the origin and then run in the direction of the coordinate triple [*xyz*], which is placed in square brackets

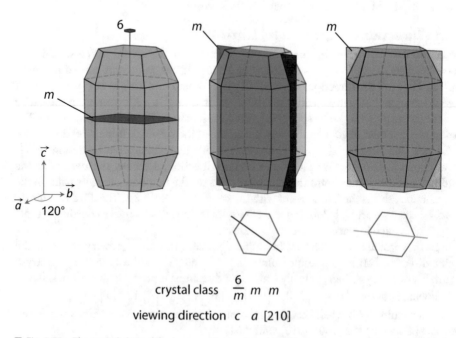

crystal class $\dfrac{6}{m}\,m\,m$

viewing direction c a [210]

Fig. 3.33 The crystal class of the hexagonal wine barrel is 6/*mmm*

lar to it there is a mirror plane m, which results in $6/m$. Along the a direction, there is another mirror plane m, and also along the direction [210], a mirror plane m appears. This results in the crystal class of $6/mmm$, which is called dihexagonal-dipyramidal according to its geometric form. All symmetry elements are shown in ◨ Fig. 3.33. However, we also know that – according to the symmetry rule – two perpendicularly oriented mirror planes generate a two-fold axis of rotation. For this reason, there are two additional two-fold axes of rotation, which are shown in ◨ Fig. 3.34. The full notation for this crystal class is therefore $\dfrac{6}{m}\dfrac{2}{m}\dfrac{2}{m}$.

Covellite (copper sulfide, CuS), for example, crystallizes in this crystal class. However, in nature the occurrence of well-defined crystal shapes is rare; more often colored, fine-grained crusts or powdery deposits on other minerals are found (◨ Fig. 3.35). Covellite was the first natural mineral at which the phenomenon of superconductivity was discovered, although at pretty low temperatures below 1.65 Kelvin [3]. Other compounds belonging to this crystal class are magnesium, graphite (see also ▶ Chap. 7), and nickeline (nickel arsenide, NiAs).

◨ **Fig. 3.34** In addition to the six-fold axis of rotation, there are two sets of two-fold axes of rotation, of which only one of each set is drawn. One is running directly through two opposite edges (green), another through two opposite face centers (orange). Equivalent axes of rotation are obtained by rotating by 60° in the (a,b) plane

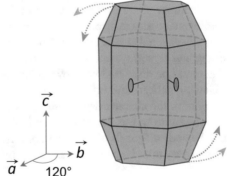

◨ **Fig. 3.35** Covellite mineral from the Leonard mine, Butte, Montana, USA, size $7.5 \times 5 \times 4$ cm. (Didier Descouens – CC-BY 3.0)

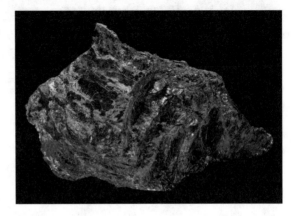

3

3.3.3   Relationship Between Crystal System and Crystal Class

Of course, the 32 crystal classes are not randomly assigned to a certain crystal system. In ▢ Table 3.3 an overview of all 32 crystallographic point groups is given, both in long and short notation, and how they are distributed over the seven crystal systems. In addition, the viewing directions for each crystal system are given, the number of rotation axes of different order, the number of mirror planes, and whether this class has an inversion center.

Some entries in the table should be discussed and commented on briefly.

In the triclinic crystal system there are only two crystal classes, 1 and $\bar{1}$, i.e., the classes in which either no symmetry element is present or only an inversion center.

In the monoclinic crystal system, there are three different crystal classes: either there is a two-fold axis of rotation or a mirror plane or a two-fold axis of rotation and a mirror plane perpendicular to it, but always in only one direction, namely, defined by convention, in the b direction.

The characteristic feature of the trigonal crystal system is that there is exactly *one* three-fold axis of rotation. You should pay careful attention to whether there is an additional mirror plane perpendicular to this three-fold axis. If this is the case, a six-fold rotoinversion axis $\bar{6}$ is present, and the structure no longer belongs to the trigonal but to the hexagonal crystal system. Conversely, not every crystal shape that belongs to the hexagonal crystal system must necessarily have a six-fold axis of rotation! This is a fact that could be discussed controversially, and some aspects of it are summarized in the following background information.

Finally, the characteristic feature of the cubic crystal system is not a four-fold rotational symmetry, but the presence of four independent three-fold axes of rotation; these are the body diagonals of each two opposite corner points. All classes in the cubic crystal system have this characteristic, as can be inferred from ▢ Table 3.3.

Classifying Crystal Classes into Crystal Systems

In former times, the classification of crystal classes or their belonging to a crystal system was based *exclusively* on the morphology of crystals. All crystals with a three-fold axis of rotation belonged to the trigonal system, and all crystals with a six-fold axis of rotation belonged to the hexagonal crystal system. Looking at ▢ Table 3.3, when looking at the trigonal and hexagonal system, it is noticeable that although all crystal classes of the trigonal system have a three-fold rotation axis, two of the crystal classes of the hexagonal system do not have a six-fold axis of rotation, namely, the crystal classes $\bar{6}$ and $\bar{6}m2$. They only have a three-fold axis of rotation, because a six-fold rotoinversion axis leads to a crystal shape with a three-fold axis of rotation. This is also reflected in the geometric names of the classes (trigonal-dipyramidal and ditrigonal-dipyramidal). It would therefore also be justified to include these two classes in the trigonal system. However, the modern assignment of crystal classes to crystal systems is based on the criterion of the order of (simple) axes of rotation *or* rotoinversion axes: all crystal classes that have a six-fold axis of rotation or a six-fold rotoinversion axis belong to the hexagonal crystal system. The advantage of

Table 3.3 Overview of all 32 crystallographic point groups

No.	Crystal system	Name	Long symbol	Short symbol	Viewing directions	2	3	4	6	m	$\bar{1}$?
1	Triclinic	Triclinic-pedial[a]	1	1	None	–	–	–	–	–	No
2		Triclinic-pinacoidal	$\bar{1}$	$\bar{1}$		–	–	–	–	–	Yes
3	Monoclinic	Monoclinic-sphenoidal[b]	1 2 1	2	b	1	–	–	–	–	No
4		Monoclinic-domatic[c]	1 m 1	m		–	–	–	–	1	No
5		Monoclinic-prismatic	$1\,\frac{2}{m}\,1$	$\frac{2}{m}$		1	–	–	–	1	Yes
6	Orthorhombic	Orthorhombic-disphenoidal	2 2 2	2 2 2	a, b, c	3	–	–	–	–	No
7		Orthorhombic-pyramidal	m m 2	m m 2		1	–	–	–	2	No
8		Orthorhombic-dipyramidal	$\frac{2\;2\;2}{m\,m\,m}$	m m m		3	–	–	–	3	Yes

(continued)

◻ Table 3.3   (continued)

No.	Crystal system	Name	Long symbol	Short symbol	Viewing directions	2	3	4	6	m	$\bar{1}$?
9	Tetragonal	Tetragonal-pyramidal	4	4	c, a, [110]	–	–	1	–	–	No
10		Tetragonal-disphenoidal	$\bar{4}$	$\bar{4}$		1	–	–	–	–	No
11		Tetragonal-dipyramidal	$\dfrac{4}{m}$	$\dfrac{4}{m}$		–	–	1	–	1	Yes
12		Tetragonal-trapezohedral	4 2 2	4 2 2		4	–	1	–	–	No
13		Ditetragonal-pyramidal	4 m m	4 m m		–	–	1	–	4	No
14		Tetragonal-scalenohedral[d]	$\bar{4}$ 2 m	$\bar{4}$ 2 m		3	–	–	–	2	No
15		Ditetragonal-dipyramidal	$\dfrac{4\ 2\ 2}{m\ m\ m}$	$\dfrac{4}{m}$ m m		4	–	1	–	5	Yes
16	Trigonal	Trigonal-pyramidal	3	3	c, a	–	1	–	–	–	No
17		Rhombohedral	$\bar{3}$	$\bar{3}$		–	1	–	–	–	Yes
18		Trigonal-trapezohedral	3 2 1	3 2		3	1	–	–	–	No
19		Ditrigonal-pyramidal	3 m 1	3 m		–	1	–	–	3	No
20		Ditrigonal-scalenohedral	$\bar{3}\dfrac{2}{m}$ 1	$\bar{3}$ m		3	1	–	–	3	Yes

No.	Crystal system	Name	Long symbol	Short symbol	Viewing directions	2	3	4	6	m	1̄ ?
21	Hexagonal	Hexagonal-pyramidal	6	6	$c, a,$ [210]	–	–	–	1	–	No
22		Trigonal-dipyramidal	$\bar{6}$	$\bar{6}$		–	1	–	–	m	No
23		Hexagonal-dipyramidal	$\dfrac{6}{m}$	$\dfrac{6}{m}$		–	–	–	1	1	Yes
24		Hexagonal-trapezohedral	622	622		6	–	–	1	–	No
25		Dihexagonal-pyramidal	$6\,mm$	$6\,mm$		–	–	–	1	6	No
26		Ditrigonal-dipyramidal	$\bar{6}\,m\,2$	$\bar{6}\,m\,2$		3	1	–	–	4	No
27		Dihexagonal-dipyramidal	$\dfrac{6\ 2\ 2}{m\,m\,m}$	$\dfrac{6}{m}\,mm$		6	–	–	1	7	Yes
28	Cubic	Tetartoidal	$2\,3$	$2\,3$	$a,$ [111], [110]	3	4	–	–	–	No
29		Diploidal	$\dfrac{2}{m}\,\bar{3}$	$m\,\bar{3}$		3	4	–	–	3	Yes
30		Gyroidal	$4\,3\,2$	$4\,3\,2$		6	4	3	–	–	No
31		Hextetrahedral	$\bar{4}\,3\,m$	$\bar{4}\,3\,m$		3	4	–	–	6	No
32		Hexoctahedral	$\dfrac{4\ 2}{m}\,\bar{3}\,m$	$m\,\bar{3}\,m$		6	4	3	–	9	Yes

[a]gr. *pedion* = basal face; this is a crystal shape in which each type of face is represented only once.

[b]gr. *sphenoid* = wedge; denotes a pair of faces, a dihedron, in which the two faces can be transferred into each other by a two-fold axis of rotation

[c]gr. *doma* = house; denotes a dihedron in which the two faces can be transferred into each other by a mirror plane

[d]gr. *scalenos* = limping, uneven; a scalenohedron denotes a polyhedron delimited by uneven triangles

3

this consideration may not be immediately apparent, but it arises from at least two points of view:

1. The first aspect is that the systematization should be as consistent as possible. Look at the classes of the tetragonal crystal system: here too, not all crystal classes have a four-fold axis of rotation. Two of the classes have only one two-fold axis of rotation. If only the criterion of the order of the rotation axes is to be used for classification, these classes should belong to the orthorhombic system. However, the presence of a two-fold axis of rotation is not a *qualifying characteristic*; keep in mind that the symmetry results in a restriction of the metric of the axis system: both a four-fold axis of rotation and a four-fold rotoinversion axis require a square base face of the unit cell ($a = b$) as well as an orthogonality of the three lattice vectors ($\alpha = \beta = \gamma = 90°$). In the orthorhombic system, the first condition is not necessarily met.

2. Classification according to the order of the axes, regardless of whether it specifies a pure rotational symmetry or a coupled rotoinversion, has a further advantage, which you can infer by inspecting the classes of the trigonal crystal system: the characteristic feature of the trigonal crystal system is that symmetry only appears in two directions. If you add the classes with a six-fold rotoinversion axis to the trigonal crystal system, you would need a third viewing direction.

3.3.4 Poster of Crystal Classes, Folding Paper Crystal Shapes, and Interactive Crystal Class Representations

Perhaps some of you are enthusiastic collectors of stones, minerals, and crystals. Maybe this book will inspire you to start building up a collection. How about you start classifying and photographing your crystal treasures, which you own or will own in the near future, according to their crystal class, and stick the photos on a visually appealing poster, in the sense of a Panini sticker collectible of the NBA or NFL? On this poster (□ Fig. 3.36), which you can download on the accompanying page of the book on the Internet ▶ https://crystalsymmetry.wordpress.com/textbook/ in the size DIN A0 (33.1 × 46.8 inches), an ideal type of crystal shape is given for each of the 32 crystal classes. The shapes of your real crystals will presumably deviate to a greater or lesser extent from them, but that is precisely what makes them so attractive. Please also note that there are not only one, but several ideal-typical shapes per crystal class (keyword Tracht! See ▶ Sect. 2.1.1).

For those of you who would like to gain a better idea of the shape of prototypical crystal classes, we have also placed corresponding cut-out sheets on the accompanying page of the book on the Internet for all 32 shapes, from which you can create corresponding paper models. The shapes and colors of the paper models are based on real existing finds of certain minerals, which are also indicated on the respective sheets.

Furthermore, the author would like to draw your attention to a great project of the University of Barcelona [4], in which interactive PDF files were created for all 32 crystal classes, showing 3D objects of corresponding symmetry, which can be freely rotated and

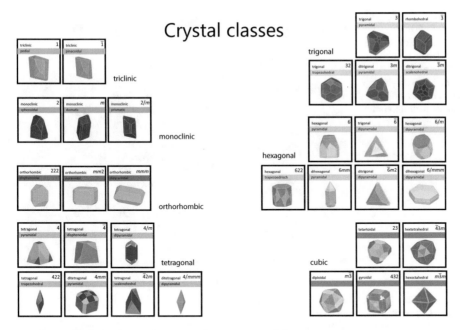

Fig. 3.36 Miniature view of the poster for the 32 crystal classes

superimposed with the symmetry elements of the class (■ Fig. 3.37). The PDF files can be downloaded free of charge at the following URL: ► https://www.uab.cat/web/la-divulgacio/grups-puntuals-de-simetria-1345664584325.html.

3.3.5 Crystal Shapes with VESTA

There are some programs that specialize in the creation and reproduction of crystal morphologies, such as SHAPE [5], WinXMorph™ [6], or KrystalShaper [7]. With the latter, the cut-out sheets for the 32 crystal classes have also been created. However, the freeware program VESTA is also able to generate crystal morphologies, which you have already got to know in ► Chap. 2 to create and view crystal structures. Here we will try to model the hexagonal wine barrel from ■ Fig. 3.31. It represents a deepening of the subject of Miller indices, because we have to consider carefully which Miller indices or face symbols the faces carry that have to be drawn.

Please first download the CIF file of the element magnesium provided on the accompanying page of the book in the web and load it into VESTA or enter the following structural data in VESTA by hand:

- Space group: $P6_3/mcm$
- Lattice parameters: $a = b = 3.19405$ Å, $c = 5.17198$ Å, $\alpha = \beta = 90°$, $\gamma = 120°$
- Fractional coordinates: Mg, $x = 0.333333$, $y = 0.666667$, $z = 0.75$

Now we have to create the barrel. We have to specify three principally different crystal faces (■ Fig. 3.38) and have to consider which Miller indices they have. The remaining faces are automatically generated by the corresponding symmetry. Think about how to specify the three faces, which are marked with (1), (2), and (3) in the illustration!

Fig. 3.37 Example of an interactive PDF document to view the crystal classes, here the crystal class 6/m, visualized by a polyhedron of corresponding symmetry with superimposed point symmetry elements

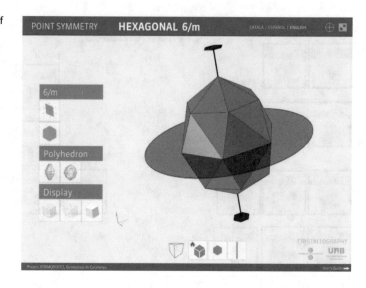

Fig. 3.38 To reproduce the crystal morphology of the hexagonal barrel, it is necessary to specify the three colored and numbered faces with their Miller indices

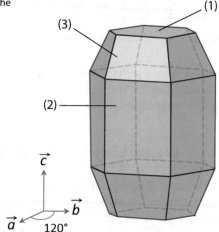

The solution:

1. The horizontal face is oriented parallel to the (a,b) plane and intersects only the c axis. So, the Miller index is what? Correct – (001)!
2. One of the vertical, rectangular boundary faces runs parallel to the (b,c) plane and intersects only the a axis, hence the Miller index (100).
3. And finally, we need a face that connects the side faces with the upper and lower faces, respectively. This intersects the a and c axis, but not the b axis, which is why the Miller index is (101).

Now we just have to take into account that we do not want to create a barrel that is only one unit cell in size. We want to create the outer boundaries of a macroscopic crystal, which may contain a huge number of unit cells. However, keep in mind that all the lattice planes of a set of lattice planes carry the same Miller indices! For this reason, it is possible to specify in

VESTA how far the corresponding lattice plane should be from the origin, either directly as a distance or in multiples of the d value of the corresponding set of lattice planes.

And now we can start with our first crystal shape.

In the first step, select *Edit – Edit Data – Crystal Shape...* from the menu and click on *New* to generate new faces. Then enter the corresponding values for the Miller indices and the distances to the origin, namely, for the face (001) 15 Å, for (100) 8 Å, and for (101) 11 Å. If a value is not accepted, press the TAB key to jump to the next field. Make sure that the *Apply symmetry operation* checkbox is checked. It is also possible to set the shade and transparency values in the line "Color (RGBA)," either numerically or by clicking on the color field (which is white in the default setting), and select a shade or click on the transparent field and move the slider according to the desired translucency level. Your program window should now look like the one shown in ◻ Fig. 3.39.

◻ **Fig. 3.39** Entering the corresponding face specification in VESTA ...

◻ **Fig. 3.40**   ...and their result

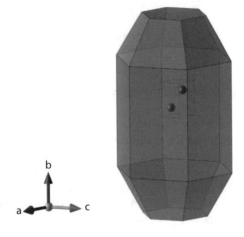

3

Fig. 3.41 The boundaries ensure that only atoms within the shape are shown

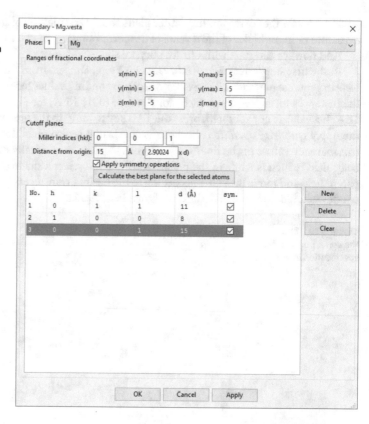

Fig. 3.42 The completed nano wine barrel, filled with magnesium atoms

After a click on the button *Apply* and *OK*, you should see something like in **Fig. 3.40**.

So far, so good, but it would still be nice if not only the two original atoms could be seen, but the whole barrel would be filled with atoms. This can be achieved by visualizing several unit cells at the same time and setting appropriate *Cutoff planes*. Under the mostly visible tab *Style*, select the button *Boundary...* and use for each cell direction −5 and 5 for the minimum and maximum value in the fields *Ranges of fractional coordinates*. Now you can use the *New* button to specify three cutoff planes, with exactly the same values as above when setting the boundary faces of the outer shape (**Fig. 3.41**). Then all the atoms should be shown up to the outer boundary faces, and all those that would protrude beyond them will be cut off. Your barrel, including all magnesium atoms, is completed (◻ Fig. 3.42)!

We will now leave the external symmetry of the crystals and their shapes. In the next chapter, we will focus on the translational symmetry of the two-dimensional plane. You will see that this has more to do with your everyday experience' than you think!

■ ■ **Summary**

In this chapter, the five point symmetry elements (identity, center of inversion, axis of rotation, mirror plane, rotoinversion axis) were introduced to describe the symmetry of macroscopic 3D bodies. The outer shape of crystals, their morphology, always corresponds to 1 of the 32 crystallographic point groups, also known as crystal classes. The point group is the systematic notation of the symmetry of this outer shape of crystals ordered according to viewing directions.

References

1. Schoenflies A (1891) Krystallsysteme und Krystallstructur. B.G. Teubner, Leizpig. https://www.archive.org/details/krystallsysteme00schogoog. Accessed 29 Sept 2019
2. Fedorov ES (1971) Symmetry of crystals. Translated from the Russian by David und Katherine Harker, New York. In: American crystallographic association monograph no. 7. American Crystallographic Association, New York
3. Di Benedetto F, Borgheresi M, Caneschi A, Chastanet G, Cipriani C, Gatteschi D, Pratesi G, Romanelli M, Sessoli R (2006) First evidence of natural superconductivity: Covellite. Eur J Min 18:283–287. https://doi.org/10.1127/0935-1221/2006/0018-0283
4. Arribas V, Casas L, Estop E, Labrodor M (2014) Interactive PDF files with embedded 3D designs as support material to study the 32 crystallographic point groups. Comput Geosci 62:53–61. https://doi.org/10.1016/j.cageo.2013.09.004
5. SHAPE V7.2, Shapesoftware, Kingsport. http://www.shapesoftware.com. Accessed 29 Sept 2019
6. WinXMorphTM V1.54, Werner Kaminsky, Department of Chemistry, University of Washington, Washington. http://cad4.cpac.washington.edu/winxmorphhome/winxmorph.htm. Accessed 29 Sept 2019
7. KrystalShaper V1.5.0 © JCrystalSoft, 2018. http://www.jcrystal.com/products/krystalshaper/. Accessed 29 Sept 2019

Symmetry in the Plane: About Wallpaper Patterns, Islamic Mosaics, Drawings from Escher, and Heterogeneous Catalysts

© Springer Nature Switzerland AG 2020
F. Hoffmann, *Introduction to Crystallography*, https://doi.org/10.1007/978-3-030-35110-6_4

4

This chapter is about finding out what the four things mentioned in the heading have in common. We will see that – as long as they are periodic patterns – all can be described with a uniform symmetry framework of the plane and can be classified into groups, depending on which specific symmetry elements are present.

In order to completely describe the symmetry of patterns in the plane, we have to deal with another symmetry element, which is not a point symmetry element but has a translational or displacement component. There are three types of symmetry that have a translational component:

- Translation vectors at which translations are executed
- Glide planes on which glide reflections (short: glides) are carried out
- Screw axes along which screw rotations are performed.

We already know the pure translations: translations by *whole* units of the primitive unit cell (from lattice point to lattice point along the three lattice vectors). Centerings must also be understood as translations, although in that case lattice points or motifs are not shifted along the lattice vectors of the unit cell and the length of the translation vectors is smaller than that of an entire unit cell length.

With glide planes and screw axes, we now want to explore two further symmetry elements, which also have a translational component that is smaller than an entire unit cell length; however, the respective symmetry operations do not consist exclusively of translations. Glide planes exist both in symmetrical patterns of the 2D plane (here they are also called glide lines) and in periodic arrangements in 3D space, whereas the presence of screw axes necessarily requires a three-dimensional periodicity. We first want to explore glide planes in two-dimensional space before discussing all the symmetry elements with translational components in space together in the next chapter.

4.1 Glide Planes

Let's first take a look at the following ◘ Fig. 4.1: all these patterns can be described by one of only 17 principally different symmetries, the so-called plane groups. There are, of course, countless possibilities to create further periodic patterns that are found in everyday life, e.g., in fabrics, clothing, wallpapers, gift wrap sheets, wall or floor tiles, and the like, because the *motifs* can be different. But with regard to the symmetry characteristics of the periodic arrangement of the motifs, there are only 17 possibilities. The reason for this is again the fact that symmetry elements cannot be combined arbitrarily with each other, because not every combination is compatible with the (strict) translational symmetry of a periodic pattern.

Let's take a few samples from ◘ Fig. 4.1 and examine them in more detail. The pattern shown in ◘ Fig. 4.2 obviously consists of a five-pointed star, which is repeated in a simple way, and the primitive unit cell is easy to locate. And apart from the mirror planes that run along a single direction, no further symmetry elements are present.[1]

---

1   Some of you may think that there are also two additional axes of rotation in the plane. This is not the case, since if the corresponding symmetry operation was carried out the motif would then temporarily leave the plane, even if it were again *in* the plane at the end of such an operation. Such operations violate the dimensionality of the object, i.e., the dimensionality of the plane, and are therefore "forbidden." A symmetry operation of the plane must remain completely *in* the plane.

Fig. 4.1 A selection of 2D periodic patterns that can be described with symmetry groups of the plane, i.e., plane groups

Fig. 4.2 An infinite pattern with a five-pointed star as motif, which is primitively arranged (the unit cell is superimposed in cyan). Apart from two mirror planes (one at the outer borders of the unit cell and one in the center of the cell, drawn in red), there is no other symmetry element

The pattern with the red flowers on a green background (**Fig. 4.3**) can be described by a centered cell, which also has only mirror symmetry in one direction. Similarly, the pattern in **Fig. 4.4** can be described with a centered unit cell, but here two perpendicularly oriented mirror planes are present. By now you know that this means that there must also be a two-fold axis of rotation. There are also patterns with higher rotational symmetry, e.g., the pattern shown in **Fig. 4.5a** shows six-fold rotational symmetry; in the pattern shown in **Fig. 4.5b**, six-fold rotational symmetry is combined with mirror symmetry.

So far nothing new. But if we look at the pattern of **Fig. 4.6**, then there is another type of symmetry besides mirror symmetry, which has not been dealt with so far: the motif (which itself is mirror-symmetrical) is not only reflected or only shifted but reflected *and* shifted simultaneously. For example, look at the pale light-blue, red-framed, fanlike arranged, pointed shapes. It will turn out that such arrangements of motifs can be described with glide reflections!

4

Fig. 4.3 This pattern of a Persian tapestry can be described by a centered unit cell in which mirror symmetry is present in one direction

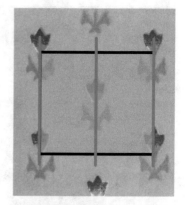

Fig. 4.4 In this Persian tapestry, a pattern was created that can be described again by a centered unit cell. In contrast to the variant in Fig. 4.3, however, the motif was chosen in such a way that mirror symmetry occurs in two directions

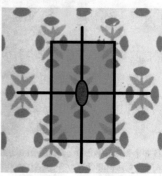

Fig. 4.5 (a) Persian ornament with (not quite perfect) six-fold rotational symmetry and (b) Byzantine pattern in which six-fold rotational symmetry is combined with two mirror planes

(a)   **(b)**

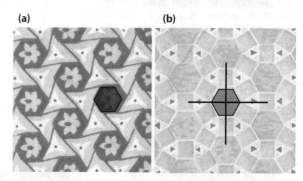

Fig. 4.6 Ancient Egyptian pattern on the ceiling of a tomb in which, in addition to mirror planes (in cyan), glide planes (black, dashed) are also present. The unit cell is superimposed as orange-shaded area

4.1.1 What Is a Glide Plane?

To deduce what a glide plane is, let's look at three different patterns of footprints, which suggest different ways of walking. If we were to move barefoot on one leg bouncing on a beach, we would leave the track shown in ■ Fig. 4.7. It is easy for us to recognize the repetitive motif, define a unit cell, and draw the translational vector that runs from one lattice point to the next.

If we jumped like a kangaroo, we would see a pattern shown in ■ Fig. 4.8 – again a simple, repeating pattern, but now combined with mirror symmetry.

If, on the other hand, we walk like a normal pedestrian – which is the more common way of walking – we would leave in the sand the pattern of footprints shown in ■ Fig. 4.9. On each side of the right or left footprints you can see the translation symmetry in whole units of the unit cell, but the two feet themselves are also related to each other in symmetry, they are mirror images! However, the two mirror-image feet are not at the same height, but are arranged staggered relative to each other, exactly half the length of a unit cell – and exactly this symmetry relationship can be described by means of a glide plane or here in the two-dimensional case with a glide line.

A glide reflection is thus a coupled symmetry operation (as well as the rotary reflections and rotoinversions described in ▶ Chap. 3), in which two operations are carried out directly one after the other and necessarily always together: first, a reflection and then a translation, usually by one half of the unit cell dimension. The glide reflection plane is

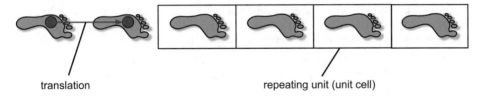

translation   repeating unit (unit cell)

■ **Fig. 4.7** Footprints that are created when we move on a beach by jumping on one leg

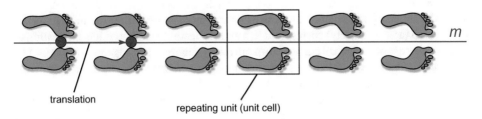

m

translation   repeating unit (unit cell)

■ **Fig. 4.8** Footprint patterns, if we were jumping on a beach like a kangaroo

■ **Fig. 4.9** This footprint pattern contains a glide line represented by a dashed line and symbolized by the small letter *g*

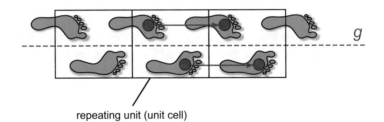

g

repeating unit (unit cell)

◘ Fig. 4.10   A schematically drawn eight with coxswain. The rowers can be mapped onto each other via the combined symmetry operation of a glide reflection. (Reproduced, original in [1], courtesy of © IUCr, ► http://www.iucr.org)

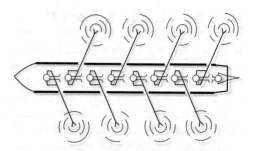

4

◘ Fig. 4.11   Some specimens of the Kaiser's spotted newt (*Neurergus kaiseri*) have a pattern of white spots on their back that are arranged approximately according to a glide plane. In this specimen, the pattern changes to a simple mirror symmetry on the tail. (Photo courtesy of Dr. Richard Bartlett)

symbolized graphically with a dashed line and abbreviated with the letter *g*. All of the footprints in ◘ Fig. 4.9 can be created from only one individual motif by applying a glide reflection: mirroring – translation by 1/2 – mirroring, etc.

An example of another macroscopic thing that reveals glide mirror motifs is an eight-seat rowing boat (◘ Fig. 4.10). However, this "crystal" would consist of only four unit cells, which would be a rather small crystal.

Furthermore, we find glide reflection motifs in coat patterns of some animal species. For example, on some (but by no means all) specimens of the Zagros newt (lat. *Neurergus kaiseri*, a species threatened with extinction, which belongs to the order of the caudates and to the family of true salamanders), an arrangement of whitish spots can be seen on the back, which (approximately) correspond to a glide reflection symmetry. It is interesting to note that these spots in the tail change into a mirror-symmetrical arrangement (◘ Fig. 4.11). However, there are also individuals among the Zagros newt who show a purely mirror-symmetrical arrangement of these spots on their backs.

Compared to these simple glide planes at periodic patterns of all kinds, crystals also contain more complicated types of glide planes, which we will discuss more deeply in ► Chap. 5. For now, we remain in the two-dimensional plane and consider the systematics and importance of the symmetry of the plane.

4.2   The Five Bravais Lattices of the Plane, the 17 Plane Groups and Their Significance for Culture and Technology

With the four symmetry operations of the plane – translation, rotation, reflection, and glide reflection – all conceivable periodic patterns of the plane can be generated. From the standpoint of pure translation, they are based on the five Bravais lattices of the

133

4

4.2 · The Five Bravais Lattices of the Plane, the 17 Plane Groups ...

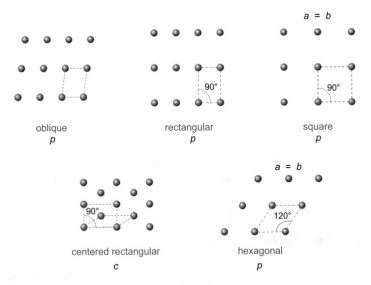

◘ Fig. 4.12 The five Bravais lattices of the plane. Underneath the lattices the centering symbol (p = primitive, c = centered) is given. For the oblique lattice, there are no restrictions regarding the lattice parameters; for the rectangular, square, and rectangular-centered lattice the angle between the lattice vectors must be 90°. For the hexagonal lattice, the angle has to be 120°. For the square and hexagonal lattice, there is an additional restriction for the lengths of the lattice vectors ($a = b$)

plane. In his phenomenal work of 1848 [2], Auguste Bravais not only described the 14 possible lattice types of 3D space, but also examined the possible lattice point arrangements in the plane and came to the conclusion that there are five[2]: the oblique, the rectangular, the square, the rectangular-centered, and the hexagonal lattice (◘ Fig. 4.12). This means that they comprise four primitive lattices and only one centered lattice.

Just as all unit cells in 3D space represent *parallelepipeds*, which are spanned by three vectors, the unit cells in two dimensions are all *parallelograms* defined by two vectors or spanned by two vectors. Special attention must be paid to the hexagonal lattice system, because, as already explained in ► Chap. 2, representations in which an enlarged cell is drawn in the form of a hexagon are simply wrong – a honeycomb pattern with lattice points at its corners does *not* form a translation lattice (► Sect. 2.7).

The combination of pure translation with the three other symmetry elements of the plane (axis of rotation, mirror plane, glide plane) leads to a total of 17 fundamentally different symmetries in 2D space – the so-called plane groups (the term "group" originates from mathematics; we will briefly discuss this term in ► Chap. 6). These 17 possibilities of symmetry of the plane was deduced by E. Fedorov in 1891 [3], interestingly, 6 years after he had catalogued the much more complicated 230 space groups (► Chap. 5).

A plane group is nothing more than a complete description of the symmetry of a periodic, two-dimensional pattern. Sometimes the completely equivalent term ornamental group is also used, indicating that symbolic decorations on fabrics, wallpaper, or buildings can be described by the same symmetry laws. Before we take a closer look at the areas in

---

2   By the way, there would be already 64 Bravais lattices in four-dimensional space.

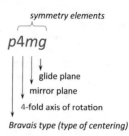

Fig. 4.13 Example for a complete notation of a plane group (see text)

4

which symmetries of the plane play a role, we will look at the nomenclature or notation and present an algorithm that makes it easier for us to determine the corresponding plane group for a given pattern.

4.2.1 Notation of Plane Groups

The (complete) symbol of a plane group always consists of four characters, an example would be *p4mg* (**Fig. 4.13**), whereby letters are always italicized and numbers are always set normally:

- The symbol always begins with the letter indicating the centering or Bravais type: *p* stands for primitive, *c* for centered.
- The second character indicates the highest order (*n*) of a possible axis of rotation; in the example shown, it is 4.
- Then there are two further symbols indicating whether there are mirror planes (*m*) or glide planes (*g*), with regard to the main (or first) and secondary viewing direction.
- If there is a mirror plane running perpendicular to one of the axes, this axis should be the main direction. This means that the mirror plane or glide plane represented by the last symbol is either parallel to the main axis or forms an angle of 180°/*n* (for *n* > 2) with it, where n, in turn, is the order of the axis of rotation of the highest order.
- If there is no axis of rotation axis or symmetry plane in the main or secondary direction, this is marked with a 1 (identity) for the sake of clarity.

The following four examples should help to clarify these rules:

- *p*311: primitive unit cell, three-fold axis of rotation, no further symmetry elements.
- *p6mm*: primitive unit cell, six-fold axis of rotation, one mirror plane perpendicular to the main direction, another mirror plane that includes an angle of 30° with the main direction and which is oriented perpendicular to the secondary direction.
- *c2mm*: rectangular-centered unit cell, two-fold axis of rotation, mirror planes both perpendicular and parallel to the main direction.
- *p4mg*: primitive unit cell, four-fold axis of rotation, mirror plane perpendicular to the main axis, glide plane at an angle of 45° to the main direction.

In addition to the long notation, there is also a short notation in which either the identity operators (e.g., *p*1*g*1 is shortened to *pg*) or those axes of rotations or mirror planes that necessarily result from the presence of other symmetry elements are omitted (e.g., *c2mm* is shortened to *cmm*). In **Table 4.1** all 17 plane groups are listed, both in their short and long notation. In **Table 4.2** the viewing directions for the crystal systems of the plane are once again explicitly indicated.

◘ **Table 4.1**   Long and short notation of all 17 plane groups

Nr.	1	2	3	4	5	6	7	8	9
Long form	p111	p1m1	p1g1	c1m1	p211	p2mm	p2mg	p2gg	c2mm
Short form	p1	pm	pg	cm	p2	pmm	pmg	pgg	cmm
Nr.	10	11	12	13	14	15	16	17	
Long form	p311	p3m1	p31m	p411	p4mm	p4mg	p611	p6mm	
Short form	p3	p3m1	p31m	p4	p4m	p4g	p6	p6m	

◘ **Table 4.2**   Viewing directions for the crystal systems of the plane

Lattice type/crystal system	1st viewing direction	2nd viewing direction
Oblique	–	–
Rectangular	a	b
Centered-rectangular	a	b
Square	a	$[110]^a$
Hexagonal	a	$[1\bar{1}0]^a$

[a]As we consider the symmetry of the plane, we would actually need only two coordinates, and the third one must be zero. However, for the sake of a uniform presentation, we always use a coordination triple

4.2.2 Algorithm for Determining Plane Groups

To anticipate, it is sometimes very difficult to assign the correct plane group to a periodic pattern. This is often due to the fact that many such patterns do not consist of separate, spatially clearly delimited motifs (e.g., point arrangements), but fill the plane completely and flow into each other. However, a little help can be provided by an algorithm in the form of a flowchart shown in ◘ Fig. 4.14, the original version of which was developed by Brian Sanderson [4]. However, it does not save us from finding the unit cell and identifying symmetry elements or answering the questions of the flowchart.

The key questions of this algorithm are as follows:

1. Is there a two-, three-, four-, or six-fold axis of rotation? Which is the rotation axis of highest order?
2. Is there a mirror plane (m)?
3. Is there a (true) glide plane (g) that does not already result from the combination of translation and reflection?
4. Is there an axis of rotation of arbitrary order (not necessarily the one with the highest order) that is *not* on a mirror?
5. Is there an axis of rotation of any order being located on a mirror?

◻ Fig. 4.14
Pattern
recognition
algorithm
according to
Brian
Sanderson
(see text).
(Redrawn with
courtesy of
Dr. Brian
J. Sanderson)

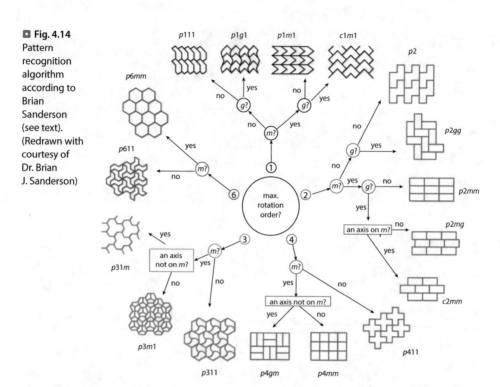

4.2.2.1 Example for Identifying the Plane Group of a Pattern

In this example, the plane group of the pattern in ◻ Fig. 4.15 should be identified. In addition, the existing symmetry elements should be superimposed on the pattern as graphical symbols – i.e., not only the presence or absence of symmetry elements should be checked, but also their *location* should be specified. This pattern does not consist of a delimited motif, and therefore, it is probably not so easy to identify and to draw the unit cell.

Following the algorithm, we should first determine whether an axis of rotation is present and, if so, what order it has. It is relatively easy to see that there is no six-, four-, or three-fold rotational axis, because a rotation by 60, 90, or 120° does not lead to an indistinguishable configuration. The question is whether there is a two-fold axis of rotation: In order to answer this, it can be helpful to restrict the area of investigation to a small part of the overall pattern and to convert the sometimes confusing colored pattern into a black and white variant (in an image editing program). It also helps to align the dominant lines of a pattern vertically or horizontally. In this way, the two-fold axis of rotation can be located (◻ Fig. 4.16).

The next level of the algorithm leads to the question whether a mirror plane is present. This can be answered relatively simply with "Yes" (◻ Fig. 4.17).

According to the algorithm, we now have to decide whether a glide plane is present – this is also the case; it runs perpendicular to the mirror plane (see ◻ Fig. 4.18). And the last question in the flowchart can also be answered immediately, namely, whether there is a (any) axis of rotation that lies on a mirror plane: this cannot be the case because the rotation axis of highest order was a two-fold axis that does not lie on the mirror plane and there are no axes of higher order. This determines the plane group to be *p2mg*. Finally, it is important to correctly define the unit cell boundaries and the position of the symmetry elements. Since a

137

4

4.2 · The Five Bravais Lattices of the Plane, the 17 Plane Groups ...

◻ **Fig. 4.15** Which plane group can be used to describe the symmetry of this pattern?

◻ **Fig. 4.16** A two-fold axis of rotation of the pattern is identified

◻ **Fig. 4.17** Also one of the mirror planes is identified

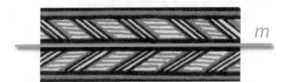

m

◻ **Fig. 4.18** The pattern belongs to the plane group *p2mg*; the unit cell is highlighted in blue, the mirror planes are shown as solid orange lines, the glide planes as dashed lines in dark blue, and the two-fold axes of rotation are marked with red and orange ellipses

p2mg

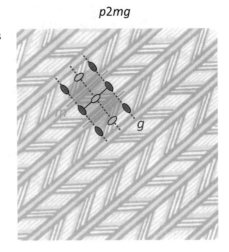

primitive Bravais lattice type is present here, we are looking for the smallest repeating unit that gives the whole pattern by simply assembling the small ones. With the symmetry elements found so far and by trial and error, finally ◻ Fig. 4.18 results in which the pattern (in b/w) is shown together with the unit cell and the correctly placed symmetry elements.

4

4.2.3 Symmetry of the Plane in Cultural Contexts

We encounter two-dimensional periodic patterns almost everywhere in everyday life: brickwork of clinker houses, artful exterior façade designs on certain buildings, pavements, parquet laminate floors, whose tiles are laid in a specific way, wall and/or floor tiles in the bathroom or kitchen, which have square, rectangular, hexagonal, monochrome, or two-tone pattern in the form of a chessboard; but also wallpaper or wrapping paper sheets, fabrics, textiles – everywhere you can find plane symmetry! Funnily enough, the term *wallpaper group* is actually more widespread in the English-speaking world than the more formal term *plane symmetry group*.

4.2.3.1 Brickwork Structures

Even in the case of seemingly more mundane things, such as the way in which bricks are put together to form a masonry bond, symmetry can be decisive – and not only for the aesthetic sensation when looking at them, but also for the stability of the masonry! Not that symmetry is the reason for the stability, because symmetry only describes the kind of arrangement according to geometric aspects, but rather in the sense that different brick assemblies differ with regard to their stability, and this is also reflected in the symmetry. The first thing a bricklayer apprentice learns – if he didn't already know – is that you don't put two rows of bricks directly on top of each other. Otherwise there would be predetermined breaking points along the mortar or the joints between the bricks. So, the first measure to increase the strength of a masonry is that every second row is installed offset by half a brick length – and the result is a glide plane motif in the masonry!

In terms of architectural history, it is extremely interesting to see in which cultural areas which kind of masonry (assemblies of bricks) have been established. There are bonds where bricks are used either only in lateral (headers) or only in longitudinal direction (stretchers), and there are those where both orientations are used. Depending on the type of spatial displacement from course to course (row to row) and the number of assemblies of similarly laid bricks at certain repeating locations, there is a very wide variety of bond patterns. A distinction is made, for example, between simple stretcher (bricks only laid in the longitudinal direction) and header bonds (bricks used in the transverse direction only) and between block and cross bonds (both are mixed forms). For exterior walls there are a number of specific mixed forms, the so-called decorative bonds, which have their own names, Monk, Flemish, Sussex, English, American, or Scottish bond, to name but a few. ◻ Figure 4.19 shows schematically some of the bonds. Are you able to assign these patterns to their corresponding plane groups?

4.2.3.2 Islamic Art and the Alhambra

The decoration of objects with symmetrical patterns, e.g., in the form of ornaments, is probably as old as mankind. This is well documented especially for the early advanced civilizations, e.g., ancient Egypt (approx. 4000 BC) and China (approx. 2000 BC). The Moors achieved a particular peak of perfection in the handling and production of symmetrical patterns that completely fill the plane. Moorish architecture is the articulated Islamic architecture of North Africa and parts of Spain and Portugal (Al Andalus), where the Andalusians (Moors) were dominant between 711 and 1492. Islamic art, which developed from the tenth century onward and was to reach its zenith in the fourteenth century, is particularly distinguished by its masterly handling of designing and realizing highly

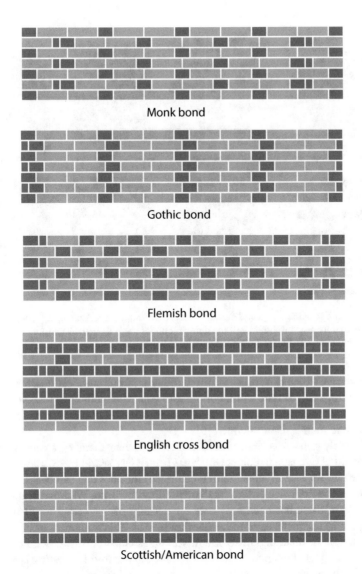

Monk bond

Gothic bond

Flemish bond

English cross bond

Scottish/American bond

▣ Fig. 4.19 A selection of masonry decorative bonds. Two of the headers (plus mortar), i.e., bricks that are incorporated in transversal direction (brown), have exactly the length of one brick that is attached in longitudinal direction (stretcher, orange). At the lateral ends of the walls often three-quarter stones or extra "ornamental stones" (1/4 width, blue) are used. The Monk bond has two stretchers between every header with the headers centered over the perpend between the two stretchers in the course below. In one of the variants of the Gothic bonds, there are also two stretchers between every header, but the headers are centered over the perpend between a stretcher and a binder in the course below and above so that they form a zigzag pattern in vertical direction. In the Flemish bond, stretchers and headers strictly alternate. The offset from layer to layer is such that a header lies exactly in the middle of the stretchers of the course above and below. In a simple (English) cross bond, there are alternating stretching and heading courses, with the headers centered over the midpoint of the stretchers. Finally, in the case of the American or Scottish bond, a varying number (5–7) of pure stretcher courses, each offset by half a brick from each other, are enclosed in a course consisting only of headers

4

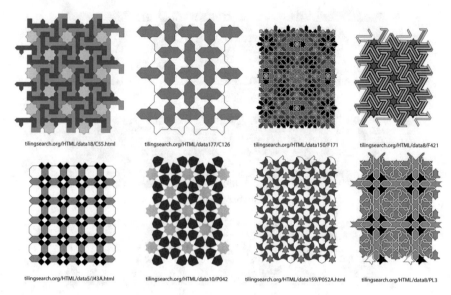

tilingsearch.org/HTML/data18/C55.html   tilingsearch.org/HTML/data177/C126   tilingsearch.org/HTML/data150/F171   tilingsearch.org/HTML/data8/F421

tilingsearch.org/HTML/data5/J43A.html   tilingsearch.org/HTML/data10/P042   tilingsearch.org/HTML/data159/P052A.html   tilingsearch.org/HTML/data8/PL3

◘ Fig. 4.20 Some selected tilings and mosaics that can be found in the Alhambra. If you want to get a comprehensive overview, you should visit the database under the URL ▶ www.tilingsearch.org in the Internet. (Reproduced with the kind permission of Brian Wichmann)

complex patterns, which can still be admired today at numerous locations, for example, in the form of tilings and mosaic work. For those of you, who would like to explore the symmetry of Islamic geometric patterns more intensively, the book by Syed Jan Abas and Amer Shaker Salman is recommended [5].

A particularly impressive example of Moorish art is the Alhambra, a palatial fortress on the Sabikah hill of Granada in Spain, which has been a UN World Heritage Site since 1984. ◘ Figure 4.20 shows a small selection of (vectorized) mosaics that can be seen there. They are taken from the extensive online database of tilings by Brian Wichmann and Tony Lee.

Interestingly, there has been a dispute for decades over whether all of the 17 plane groups appear in the patterns of the Alhambra [6–11]. This is another indication of how difficult it can be to unambiguously identify plane groups. For those who created these patterns – the master craftsmen of this epoch – the answer to this question is as meaningless as it is for those of us who simply enjoy the beauty and diversity of these mosaic works, because the mathematical framework for describing the symmetry of the plane was developed only five centuries later.

4.2.3.3 Maurits Cornelis Escher

In 1922, the famous Dutch artist M.C. Escher visited the Alhambra for the first time in his life, a visit that should determine much of his later life's work. Escher was fascinated by the variety of decorative mosaic works and was at the same time somewhat surprised that in these highly symmetrical, aesthetic patterns, there were no signs of any human or animal, not even plant-like forms.[3] He considered this at the same time to be a strength and weakness of the mosaics. In the following years, Escher contemplated from time to time, how he

---

3   The fact that no people and animals were depicted could have something to do with the alleged ban on icons/images in Islam, see also: ▶ https://en.wikipedia.org/wiki/Aniconism_in_Islam

would be able to design stronger organic but still congruent figures that would fill the plane completely, i.e., leave no gaps. In 1926, he succeeded in doing so for the first time with a lionlike figure that he also showed in two exhibitions, but which – to his disappointment – received little attention. For 10 years, those figures remained rather a hobby for Escher. But with his second visit to the Alhambra in May 1936, together with his wife Giuliaetta ("Jetta") Umiker (also a gifted and ambitious artist), his interest in it was awakened to a new life – this time in a sustainable manner! Within 3 days, they produced numerous sketches of the motifs of the Alhambra. Returning home, Escher showed these motifs to his half-brother, a professor of geology, who recommended to him a list of the most important articles from crystallographic journals dealing with the symmetry of the plane, although most of them were too theoretical or too difficult, as Escher admitted. But some of them were clear and descriptive enough to drive his work forward decisively. Thus, in the following years, he created a large number of periodic 2D patterns composed of imaginative, complementary organic figures/motifs (clowns, reptiles, birds, swans, and much more) with completely different types of symmetry, some of which contain only point symmetry elements and others also glide planes. You can get a nice overview of his symmetry-related drawings at the following URL: ▶ https://www.mcescher.com/gallery/symmetry/.

Escher's impact and influence on other artists was and still is tremendous. There are many artists who are inspired by him and create drawings based on identical construction rules but with their own motifs, see, for instance, the two drawings in ◨ Figs. 4.21 and 4.22 by Regolo Bizzi.

However, this is only a small excerpt from Escher's extensive oeuvre. Equally popular are his illustrations based on optical illusions leading the viewer into believing in the possibility of construction of perpetuum mobiles. For anyone who wants to learn more about Escher's life and work – who has always struggled for recognition in the art scene but was appreciated even more by mathematicians and crystallographers – the author recommends Doris Schattschneider's splendidly equipped illustrated book *Visions of Symmetry* [12].

◨ **Fig. 4.21** A bird-like arrangement in the style of one of Escher's drawing. (With kind permission by Regolo Bizzi, ▶ https://www.facebook.com/regolo54)

■ **Fig. 4.22** A fish-and-butterfly-like arrangement in the style of one of Escher's drawing. (With kind permission by Regolo Bizzi, ▶ https://www.facebook.com/regolo54)

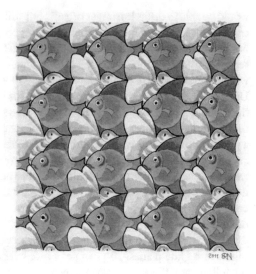

4.2.4 **Symmetry of the Plane in Technology and Relations Between 3D and 2D Bravais Lattices**

After this (hopefully entertaining) excursion into cultural spheres, we want to return to crystals. The fundamental meaning of the five Bravais lattices of the plane and the 17 plane groups results from a very simple context: although crystals are three-dimensional bodies, we have already learned in ▶ Chap. 2 that each crystal is terminated by planar faces which run parallel to certain sets of lattice planes. This means *that every crystal face can be described by a 2D Bravais lattice*. We are about to approach the domain of surface physics and chemistry – disciplines that deal with the geometry, structure, electronic properties, and adsorption of substances on the surfaces of solids. The properties at the interfaces of crystals differ considerably from those in volume; this should already be obvious because the atoms or motifs directly at the interface have a different number of neighboring atoms/motifs than the ones further inside, in the bulk state.

Crystal surfaces are of eminent importance in the area of heterogeneous catalysis, in which the catalyst of a chemical reaction is in a phase other than that of the reacting species. In technical processes, the heterogeneous catalyst is often present as a solid, while the reactants are gases or liquids.

One of the most important technical processes in which a heterogeneous catalyst is used is the Haber-Bosch process, which has been operating since 1916/1917 as a large-scale industrial process for the production of ammonia from the gaseous elements hydrogen and nitrogen and which runs according to the following reaction equation:

$$N_2 + 3H_2 \rightarrow 2NH_3$$

A catalyst mixture is used which, under the conditions of the reaction, produces the actually active catalyst species α-iron (ferrite). Ferrite is the body-centered cubic (bcc) modification of iron.

In Belgium's capital Brussels, we can admire its landmark the Atomium (■ Fig. 4.23) that shows a model of the unit cell of ferrite enlarged 165 billion times. It was constructed on the

◻ **Fig. 4.23** The Atomium, the symbol of the city of Brussels, built for the 1958 World's Fair, is a 165 billion-fold enlarged model of the body-centered cubic unit cell of alpha iron. (Reproduced with kind permission of the V.o.G. Atomium)

occasion of the World's Fair in 1958 and was to be a symbol for the nuclear age and the peaceful use of nuclear energy. It would also be justified as a monument to the role of α-Fe in the Haber-Bosch process, because ammonia is the most important raw material for mineral and nitrogen fertilizers. Estimates from 2008 assume that the nutrition of about half of the world's population is based on nitrogen fertilizer production from the Haber-Bosch process [13].[4]

The exact mechanism of the reaction to ammonia and the individual steps that take place at the catalyst surface could only be clarified many decades after the introduction of the process. Gerhard Ertl (*10.10.1936, Stuttgart, Germany) played a major role in this, and in 2007 he received the Nobel Prize for chemistry for his pioneering studies and insights into the processes of chemical reactions on solid surfaces. The greatly simplified reaction sequence is illustrated in ◻ Fig. 4.24.

Why am I telling you all this? Well, one of the most exciting discoveries of surface chemistry is that the reactivity of crystalline surfaces is extremely diverse. In concrete terms, for example, the effectiveness of the iron catalyst depends heavily on which surface is involved in the reaction, and this in turn depends on the morphology of the iron particles, i.e., which sets of lattice planes build the outermost surfaces. It has been shown that the (111) and (211) faces are by far the most reactive surfaces. But how do they look like? To this end, it is useful to visualize the correlations between the 3D Bravais lattices of a given crystalline phase and the corresponding 2D Bravais lattices of certain crystal

---

4   However, before one falls into too much admiration for Fritz Haber and Carl Bosch, one should take a closer look at their – to put it mildly – at least ambivalent biographies.

4

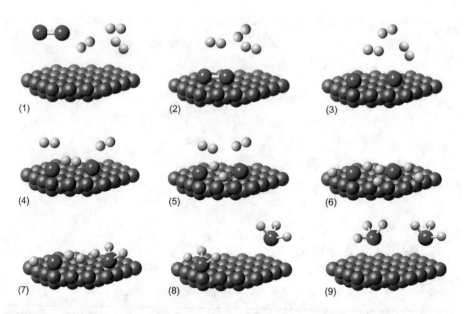

◘ Fig. 4.24 The highly simplified reaction sequence of ammonia synthesis on a Fe(111) surface (green). The steps include (1) nitrogen (N_2, blue) and hydrogen (H_2, gray) are first in the gas phase, (2) the nitrogen molecule adsorbs at the iron surface and dissociates into two single nitrogen atoms (3), (4, 5, 6) hydrogen molecules also adsorb at the surface and dissociate into hydrogen atoms, (7, 8, 9) ammonia molecules form on the surface and desorb from iron into the gas phase

faces. Since metals, i.e., pure elements, are very often used in heterogeneous catalysis, even the arrangement of the *lattice points* may correspond to the *atomic* arrangement.

The relationships are to be shown on the basis of the most important types of lattices and selected lattice planes, namely, the body-centered cubic, the face-centered cubic, and the hexagonal type. For the sake of completeness, the primitive cubic lattice type should also be discussed, although this is technically less important because there is only one single metal that crystallizes in this type of lattice, the radioactive polonium.

4.2.4.1 Primitive Cubic Metals (Only Polonium)

If we recall the shape of the primitive cubic unit cell, it is not difficult to see that the (100) face, equivalent to a lattice plane with Miller indices (100), results in the square 2D Bravais lattice type ($a = b$, angle = 90°). It is also easy to deduce that the (110) plane, i.e., the face diagonal of the cube, results in the rectangular lattice type with an aspect ratio of $\sqrt{2}:1$ and an angle of 90°. For some readers, it may be somewhat more surprising to find that the (111) surface, i.e., the space diagonal face of the primitive cube, produces the hexagonal lattice type ($a = b$, angle = 120°). Finally, the (211) surface again produces the rectangular lattice type, now with an aspect ratio of $\sqrt{3}:\sqrt{2}$. In ◘ Fig. 4.25 the relations are once again summarized graphically.

4.2.4.2 Body-Centered Cubic Metals (V, Nb, Ta, Cr, Mo, W, α-Fe, etc.)

What are the correlations when we move from the primitive cubic to the body-centered cubic-inner-centered lattice type? Just like that? Only almost! The (100) surface – logical – corresponds to the square lattice type. However, the (110) face does not result in the rect-

145   **4**

4.2 · The Five Bravais Lattices of the Plane, the 17 Plane Groups ...

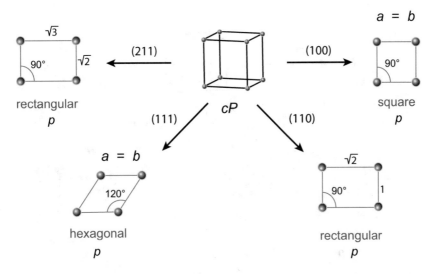

◻ Fig. 4.25 The relations of the primitive cubic Bravais lattice with the corresponding 2D Bravais lattices of selected surfaces

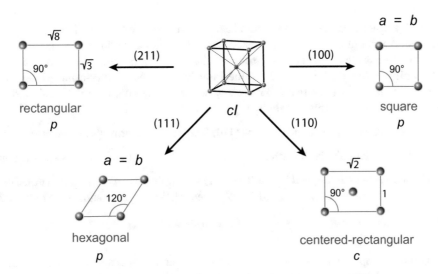

◻ Fig. 4.26 Correlation between a body-centered cubic lattice and the corresponding 2D Bravais lattices of some surfaces

angular lattice type, but in the centered-rectangular type. Identical to the primitive cubic lattice are again the cases for the (111) plane resulting in the two-dimensional hexagonal lattice and for the (211) plane resulting in a rectangular lattice, this time with an aspect ratio of $\sqrt{8} : \sqrt{3}$. In ◻ Fig. 4.26 the relationships are illustrated again.

4.2.4.3 Face-Centered Cubic Metals (Rh, Ir, Ni, Pd, Pt, Cu, Ag, Au, etc.)

The (100) surface of a face-centered cubic lattice could be described by two different 2D Bravais lattices: on the one hand by the rectangular-centered type, with the peculiarity

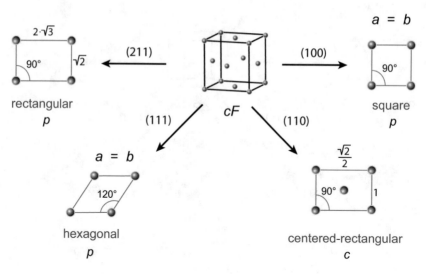

☐ **Fig. 4.27**   Relationships of the face-centered cubic Bravais lattice with the corresponding 2D Bravais lattices of selected surfaces

that in this case the lattice vectors would have the same length ($a = b$). However, we have to take into account that in two dimensions the same rules apply for setting up unit cells as in the three-dimensional case: for example, the base vectors should be as short as possible, which is not the case for the unnecessarily large rectangular-centered cell. The primitive-square lattice rotated by 45° already reflects the symmetry of the lattice and has shorter base vectors, so this possibility is preferable.

Therefore, it is easy to deduce that the (110) face – as in the primitive cubic case – again results in the rectangular lattice type, here with an aspect ratio of $1 : \dfrac{\sqrt{2}}{2}$, and a section along a (111) face reveals a hexagonal lattice. The (211) face again gives a rectangular lattice, now with an aspect ratio of $2 \times \sqrt{3} : \sqrt{2}$. The established relationships are shown in ☐ Fig. 4.27.

4.2.4.4   Metals with a Hexagonal Closest Packing (Mg, Ti, Zr, Re, Ru, Os, Co, Cd, etc.)

The two most important types of surfaces in the hexagonal are (a) the faces that run parallel to the basal surfaces of the hexagonal prism and (b) the faces that run parallel to the prism edges; these are the (001) and (100) faces. Here, the result is a hexagonal and rectangular 2D Bravais lattice, respectively, whereby no aspect ratio can be specified for the latter, since the length of the c-axis is not defined in the hexagonal system. The surfaces (110) and ($\bar{1}$10) result in rectangular lattices and the surfaces (111) as well as (211) in oblique lattices. The statements are summarized in ☐ Fig. 4.28.

Of course, the lattice type of a crystal surface is not the only aspect of heterogeneous catalysis. It should also be kept in mind that the lattice point arrangement does not have to

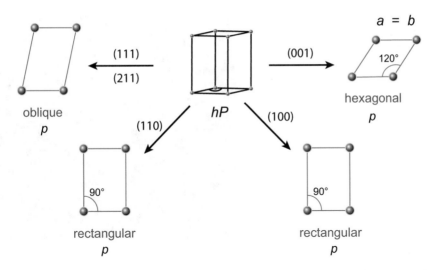

oblique
p

(111)
(211)

hP

(110)

(001)

(100)

a = b

120°

hexagonal
p

90°

rectangular
p

90°

rectangular
p

◨ **Fig. 4.28** The relationships between the primitive-hexagonal Bravais lattice and the corresponding 2D Bravais lattices of selected surfaces

correspond to the atomic arrangement for all surfaces. In addition, we have to consider the metric, i.e., how far the lattice points are apart from each other, because the particles that adsorb on a surface not only "see" the very top layer but also have contact with the layer below due to the exposed hollows of the top layer; this, of course, depends on the size of the adsorptive. Of course, electronic effects and the physicochemical nature of a given layer also play a role, which cannot be deduced from the geometry of the lattice alone. This is one of the reasons why the individual steps and the process of heterogeneous catalysis are so complicated, and the search for suitable catalysts and their optimization can be tedious.

4.3 Be Creative: Design Your Own 2D Patterns!

There are now a number of tools on the Internet that allow you to create symmetrical patterns in the plane. These are special drawing or painting programs that will run in the browser and in which you can set either a desired non-crystallographic point symmetry or a plane symmetry. These programs automatically ensure that the resulting drawing corresponds to the chosen symmetry. Try, for instance, "Kali" [14], "Escher Web Sketch" [15], "iOrnament" [16], or "Symmetry Artist" [17].

Particularly aesthetic and colorful paintings (◨ Fig. 4.29) can be created very easily with the software "Weave Silk," which is also available as a free app for the iPhone or iPad (▶ http://weavesilk.com/). A downloadable program for the Windows and Mac operating systems is "GeCla" ("Generator and Classifier"), which not only takes plane symmetry into account when drawing but is also able to analyze and classify point symmetrical or 2D periodic patterns [18].

◨ Fig. 4.29 An image created with "Weave Silk" showing superimposed patterns of different rotational and mirror symmetry

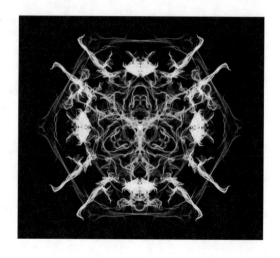

4

◨ Fig. 4.30 Over-view of the different types of symmetry that occur in closed, finite bodies (point symmetry), as well as in periodic objects in the plane and in space

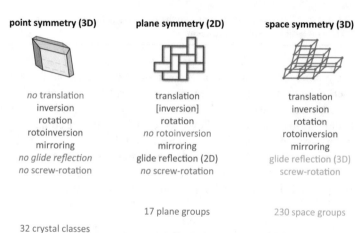

point symmetry (3D)	**plane symmetry (2D)**	**space symmetry (3D)**
no translation	translation	translation
inversion	[inversion]	inversion
rotation	rotation	rotation
rotoinversion	*no* rotoinversion	rotoinversion
mirroring	mirroring	mirroring
no glide reflection	glide reflection (2D)	glide reflection (3D)
no screw-rotation	*no* screw-rotation	screw-rotation
	17 plane groups	230 space groups
32 crystal classes		
	5 Bravais lattices	14 Bravais lattices

4.4 Intermediate Conclusion and Transition to Symmetry in Space

Up to this point of the book, we have worked out the symmetry of closed, macroscopic bodies and know the laws of symmetry of the plane. In addition, the basic features of the symmetry of space were conveyed, but two elements of symmetry – the glide plane in 3D space and the screw axes – are missing for its complete description (see the diagram in ◨ Fig. 4.30).

To describe the symmetry of the outer shapes of crystals, we only need four symmetry elements or operations, if we disregard the identity, which is inherent in each object. The specific combination of symmetry elements that are present at a given crystal leads to 32 different classes called crystallographic point groups. The reason for the limitation of the number of different crystal classes is the limitation in rotational symmetry. Interestingly, no (almost) perfectly grown single-crystals can have a rotational symmetry of the order five or of the order of more than six. What is missing at the outer symmetry of objects – crystals as well as everyday objects – is the periodicity. Periodicity cannot exist on a single, isolated object; this requires an ensemble of similar objects that are arranged regularly – at equal intervals from each other and in identical orientation – so that they form a lattice. A lattice point represents an object or motif, and the symmetry operation that maps a lattice point to an adjacent one is the translation.

All two-dimensional periodic arrays of objects in the plane are based on one of five Bravais lattices. The combination of point symmetry elements, compatible with the five Bravais lattices, lead to 17 different plane groups. In addition to translation, the glide plane or glide line has been added as a new symmetry element, also a symmetry element with a translational component. With regard to the point symmetry elements, the inversion is omitted compared to macroscopic objects, because an inversion in the 2D plane is equivalent to a rotation of 180°.

If we shift now to the symmetry of space, another symmetry operation will be added, which has a translational component, the screw rotation. Their nature – a rotation in the plane (two dimensions) and a translation perpendicular to that plane (third dimension) – necessarily requires a spatial expansion in all three dimensions. While there is only one type of glide plane in two dimensions, in three dimensions there are further possibilities to combine reflections with a translation. The number of Bravais lattices increases to 14, and the possible combinations of all point symmetry elements with the different lattice types result in 230 different types of spatial symmetry, which are described by the space groups. The development of the understanding of the complete spatial symmetry will be worked out in the following two ▶ Chaps. 5 and 6.

References

1. Dent Glasser LS (1984) Symmetry. IUCr teaching pamphlet no. 13. University College Cardiff Press for the International Union of Crystallography, p 9
2. Bravais A (1850) Mémoire sur les systèmes formés par des points distribués regulièrement sur un plan ou dans l'espace. J Ecole Polytech 19:1–128
3. Fedorov E (1891) Symmetry in the plane. Reprint in: Proceedings of the Imperial St. Petersburg Mineralogical Society. 28:245–291. (in Russian)
4. http://homepages.warwick.ac.uk/~maaac/. Accessed 29 Sept 2019
5. Abas SJ, Salman AS (1995) Symmetries of Islamic geometrical patterns. World Scientific, Singapore
6. Müller EA (1944) Gruppentheoretische und Strukturanalytische Untersuchungen der Maurischen Ornamente aus der Alhambra in Granada. Dissertation, Universität Zürich
7. Grünbaum B, Grünbaum Z, Shepard GC (1986) Symmetry in Moorish and other ornaments. Comput Math Applic 12:641–653. https://doi.org/10.1016/0898-1221(86)90416-5
8. Pérez-Gómez R (1987) The four regular mosaics missing in the Alhambra. Comput Math Applic 14:133–137. https://doi.org/10.1016/0898-1221(87)90143-X
9. Grünbaum B (2006) What symmetry groups are present in the Alhambra? Not Am Math Soc 53: 670–673
10. Blanco MF, Harris ALN (2011) Symmetry groups in the Alhambra. VisMath 13:1–42
11. Bodner BL. The planar crystallographic groups represented at the Alhambra. Proceedings of Bridges 2013: Mathematics, Music, Art, Architecture, Culture. http://archive.bridgesmathart.org/2013/bridges2013-225.pdf. Accessed 29 Sept 2019
12. Schattschneider D, Escher MC (2004) Visions of symmetry. 2nd edn. Harry N. Abrams, Inc., New York
13. Erisman JW, Sutton MA, Galloway J, Klimont Z, Winiwarter W (2008) How a century of ammonia synthesis changed the world. Nat Geosci 1:636–639. https://doi.org/10.1038/ngeo325

14. Kali. http://www.tessellations.org/software-kali-win1.shtml. Accessed 29 Sept 2019
15. Escher Web Sketch 2. http://escher.epfl.ch/index.html. Accessed 29 Sept 2019
16. iOrnament. https://imaginary.github.io/applauncher2/. Accessed 29 Sept 2019
17. Symmetry Artist. https://www.mathsisfun.com/geometry/symmetry-artist.html. Accessed 29 Sept 2019
18. GeCla 0.95. Associação Atractor. http://www.atractor.pt/mat/GeCla/index.html. Accessed 29 Sept 2019

4

Translational Symmetry Elements in Crystals and Space Groups I

© Springer Nature Switzerland AG 2020
F. Hoffmann, *Introduction to Crystallography*, https://doi.org/10.1007/978-3-030-35110-6_5

In the previous chapter, we got to know one of three symmetry elements with a translation part: the glide lines or glide planes, respectively. In this chapter, we want to discuss all symmetry elements with a translational component that can be present in crystals. This is a prerequisite for being able to fully describe the symmetry of periodic patterns of three-dimensional space. The latter is done with the space groups. But for now, let's start with the glide planes in crystals.

5.1 Glide Planes in Crystals

As a brief repetition, we should remember once again that we must distinguish between the symmetry operation and the symmetry element (Fig. 5.1): the symmetry operation is the glide reflection, while the symmetry element, i.e., the geometric element on which the operation is performed, is the glide plane.

A glide reflection is (like for example the rotary inversion) a *coupled* symmetry operation, in which two operations must be executed, necessarily belonging together: first an object is mirrored at a plane and then a translation operation is carried out, usually by one half of the unit cell length along a specified direction. A special feature of crystals is that there are also glide planes, in which the displacement component is only a quarter of the unit cell dimension.

There are a total of six different glide planes which can occur in crystals; they are called a, b, c, n, d, and e. We want to start with the three simpler glide planes a, b, and c, in which the name of the glide plane practically corresponds to the direction of the translation component: in the glide plane a, the translation takes place along a or the x axis, respectively, and so on. Then we will treat the more complicated glide planes n, d, and e. With regard to the orientation of the mirror components of the glide planes to the coordinate system, they all have in common that these planes are either at least perpendicular to an axis or even parallel to a plane spanned by two vectors of the coordinate system. This means, fortunately, we do not have to deal with glide planes where the plane is completely oblique in the unit cell. Expressed by Miller indices, at least one index $[(0kl), (h0l), (hk0)]$ or even two indices are zero $[(00l), (0k0), (h00)]$.

 Fig. 5.1 Overview of the symmetry element glide plane and the associated symmetry operation glide reflection

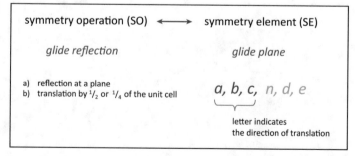

symmetry operation (SO) ⟷	symmetry element (SE)
glide reflection	*glide plane*

a) reflection at a plane
b) translation by $1/2$ or $1/4$ of the unit cell

a, b, c, n, d, e

letter indicates the direction of translation

5.1.1 The Three Simple Glide Planes *a*, *b*, and *c*

As mentioned above, the common feature of the three simple glide planes is that their names indicate the direction of the translation. Furthermore, for each of them, the translational component is exactly one half of the unit cell length.

It is common to graphically visualize glide planes and their effect on an object in the unit cell in such a way that a two-dimensional projection, a drawing plane, is viewed. With regard to the orientation of the glide plane, only two different cases have to be distinguished from each other, although three cases are actually conceivable:

1. The glide plane is parallel to the (*a*,*b*) plane.
2. The glide plane is parallel to the (*b*,*c*) plane.
3. The glide plane is parallel to the (*a*,*c*) plane.

But one possibility can be excluded for each case, namely, that in which the direction of the glide plane – expressed by the normal vector, which is perpendicular to this plane – would coincide with the direction of the translation component; this is simply not possible. For the subsequent considerations, we uniformly select the (*a*,*b*) plane as the drawing plane.

5.1.1.1 Glide Plane *a*

In ◻ Fig. 5.2 we see a gray drawing plane, which should represent the projection of the unit cell onto the (*a*,*b*) plane; the *c* axis should be perpendicular to this plane for simplicity (but this does not always have to be the case).

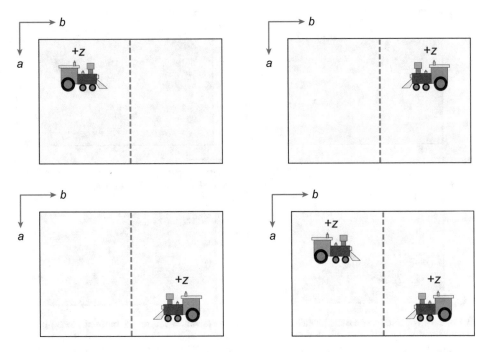

◻ **Fig. 5.2** The impact of a glide plane *a* oriented perpendicular to the drawing plane on the motif of a locomotive. The starting point is top left, and the locomotive is above the drawing plane (+z), is mirrored (top right) with no change in height (+z), and finally shifted by ½ in positive *a* direction (bottom left). Both locomotives, which are symmetry-related via the glide plane, are shown together in the lower right corner

In the first case, the glide plane should be perpendicular to the drawing plane, i.e., parallel to the (a,c) plane; in ◘ Fig. 5.2 it is drawn with a dashed line. Now we place an object into the unit cell, one whose outer form also makes the effect of a reflection noticeable – here a locomotive. If we have chosen just a simple circle or a sphere, we are able to trace the change of location caused by the reflection, but not the mirror effect on the *motif*! The locomotive should first be somewhere above the projection plane at the height $+z$ (◘ Fig. 5.2, top left). Now we want to perform the glide reflection: We must first mirror the object on the dotted plane (◘ Fig. 5.2, top right) – the fractional coordinate $+z$ will not change by that operation – and finally we have to shift the locomotive by ½ in the positive a direction (◘ Fig. 5.2, bottom left). Of course, the original locomotive does not disappear from its original place – to make this clear, in ◘ Fig. 5.2, bottom right, the two symmetry-related (connected to each other via the glide plane) locomotives are shown together in the unit cell.

The second case that has to be considered is that the glide plane a lies in the (a,b) plane, i.e., parallel to the drawing plane; for simplicity the concrete position along the c axis should be $c = 0$, meaning that the glide plane and the drawing plane coincide. We will see later that the position of the glide plane does not necessarily have to be at the height $c = 0$; often it is also at the positions $c = ½$ or $c = ¼$.

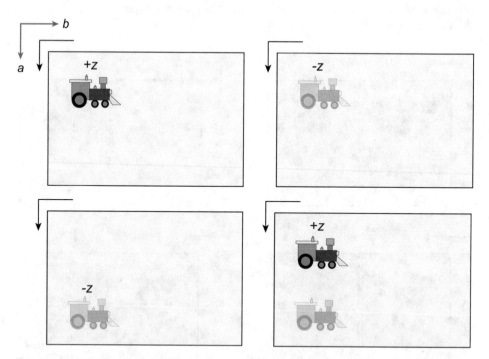

◘ **Fig. 5.3** The impact of a glide plane a, which lies in the drawing plane, on the motif of a locomotive. The locomotive should initially be above the drawing plane ($+z$) and be mirrored (top left), is then located below ($−z$) the plane of the glide plane (top right), and is finally shifted by ½ in the positive a direction (bottom left). Both symmetry-related locomotives are shown together at the bottom right

Fig. 5.4 There is no glide plane, for which its normal vector (here a) corresponds to the direction of translation

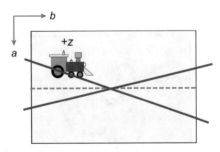

Since the position can no longer be indicated by a dashed line, an additional angle (top left in ◘ Fig. 5.3) is used, whereby an arrowhead indicates the translation direction (along a). We start with a similar scenario as in the first case and place our motif in the unit cell at a position above the glide plane, at the height $+z$. Carrying out the reflection (◘ Fig. 5.3, top right) first transports the locomotive below the mirror plane to the height $-z$ (indicated by the pale color of the locomotive), and now the translation along the a direction must be carried out; hence, the x coordinate increases by 0.5 (◘ Fig. 5.3, bottom left). In ◘ Fig. 5.3, bottom right, both locomotives, which are symmetry-related to each other via the glide plane, are once again shown together.

As already explained at the beginning of this section, the third, purely theoretically conceivable case with a glide plane parallel to the (b,c) plane does not have to be considered (◘ Fig. 5.4), since the direction of the movement that would accompany the reflection and the direction of the translation would be identical.

5.1.1.2 Glide Plane b

The glide plane b can be treated completely analogous to the glide plane a. Again, a distinction must be made between the two cases in which the mirror plane is parallel to the drawing plane or perpendicular to it. In both cases, the translation part is $+\frac{1}{2}$ in direction b, so $\frac{1}{2}$ must be added to the fractional coordinate y of the corresponding atom/motif. In the first case (glide plane perpendicular to the drawing plane), the locomotive should be above the drawing plane in the upper left corner of the unit cell (◘ Fig. 5.5, top left). After reflection (◘ Fig. 5.5, top right), a translation of $+\frac{1}{2}$ along the b direction (◘ Fig. 5.5, bottom left) is carried out. The two locomotives related to each other by this glide plane b are shown together in ◘ Fig. 5.5, bottom right.

In the second case, the glide plane is oriented parallel to the drawing plane, which in turn is indicated by an angle; an arrowhead indicates again the translation direction of that glide plane b. The locomotive should initially be somewhere at $+z$ above the mirror plane (◘ Fig. 5.6, top left), the reflection transports it by the same amount below the plane ($-z$, locomotive shown in pale colors), and, finally, the translation along one half of the unit cell along the positive b direction has to be carried out, resulting in the arrangement shown in ◘ Fig. 5.6, bottom right.

5.1.1.3 Glide Plane c

You have already understood the principle of the three simple glide planes, so you can easily complete the drawings in ◘ Fig. 5.7. For the two cases of the glide plane c that have to be considered, the initial situation is shown with a locomotive placed in the unit cell.

5

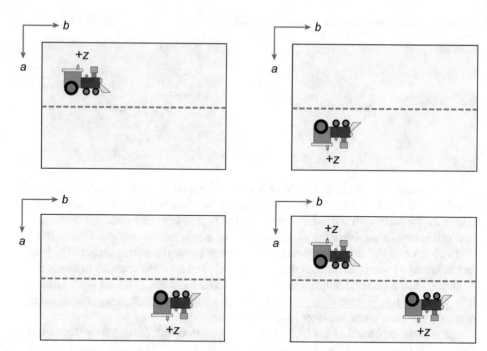

◼ **Fig. 5.5**   The impact of a glide plane *b*, which is oriented perpendicular to the drawing plane, on the motif of a locomotive. This locomotive is initially located above the mirror plane (top left), is then mirrored (top right), and is shifted by +½ in the *b* direction (bottom left). The corresponding overall picture is shown at the bottom right

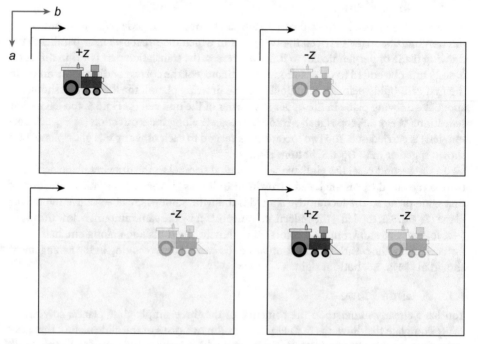

◼ **Fig. 5.6**   A glide plane *b*, located in the drawing plane, transports a locomotive, which is initially located above (+*z*) the drawing plane (top left), below the mirror plane (−*z*, pale, top right), and then moves it in the positive *b* direction by ½ (bottom left). The initial and final states are shown together in the lower right corner

Note that, here, because we have selected the (a,b) plane as the drawing plane, the glide plane is now perpendicular to the drawing plane for both cases.

5.1.2 The Three More Complicated Glide Planes *n, d*, and *e*

The three glide planes *n, d*, and *e* are, in contrast to the three simpler glide planes *a, b*, and *c*, characterized in that either the translation does not run in the direction of one of the lattice axes (*n*), that the extent of the translation is only ¼ instead of ½ of the unit cell length (*d*), or that instead of only one, two glide planes of the same orientation and location (*e*) are present, which form a so-called double glide plane.

5.1.2.1 Glide Plane *n*

The direction of translation of the glide plane *n* runs along the diagonal (*n* as in diagonal), or in other words, two translations must be carried out, each by ½ along two lattice vectors, which in sum, of course, corresponds to a translation by ½ of the diagonal.

Again, two scenarios have to be considered. In the first case, the glide plane should be oriented perpendicular to the drawing plane (◻ Fig. 5.8); a diagonal glide plane is marked with a dashed-dotted line. The locomotive should be located initially behind the drawing plane at the height *z* (locomotive in pale colors) and is then reflected at the glide plane, but the height does not change; it is still below the drawing plane and is then translated by +½ in the direction *b* and again by +½ in direction *c* so that the locomotive is finally located at the height *z* + ½. Since you are already familiar with the effect of glide planes, ◻ Fig. 5.8 only shows the initial (left) and final state (right).

In the second conceivable case, the glide plane is located parallel to or lies in the drawing plane, respectively. For clear identification, an angle is drawn next to the projection of the unit cell, in which the direction of the translation is also shown with an arrow. In the initial situation, the locomotive should lie below the drawing plane (◻ Fig. 5.9, left, pale-colored locomotive). First, the reflection will invert the *z* coordinate. Then it is translated along the diagonal of the (a,b) plane by ½ so that both the *x* and *y* coordinates increase by ½. The two motifs, which are connected by this glide plane *n*, are shown together in ◻ Fig. 5.9 on the right.

Although ◻ Fig. 5.8 and 5.9 look very similar, the effect of the two glide planes is not identical – note the different orientations of the locomotives.

5.1.2.2 Glide Plane *d*

The glide plane *d* is also called diamond-like glide plane because such a type of glide plane is actually present in the crystal structure of diamond. This is also easy to remember: *d* for diamond! In diamond-like glide planes, also a diagonal translation is performed but – unlike in glide planes *n* – only by one quarter of the unit cell dimension.

In the following scenario, a glide plane *d* will be illustrated: the glide plane should be oriented perpendicular to the drawing plane, the (a,b) plane; graphically it is symbolized by an arrow-dash line (◻ Fig. 5.10). Locomotive ① should be somewhere above the drawing layer. The reflection first transports the locomotive to the other side of the mirror, and it reverses its orientation (locomotive ②). Then a diagonal translation by ¼ takes place along the (b,c) plane, i.e., the fractional coordinates of *y* and *z* increase each by ¼. But, of course, the glide plane *d* must also be applied to the locomotive ② at its new location! This

5

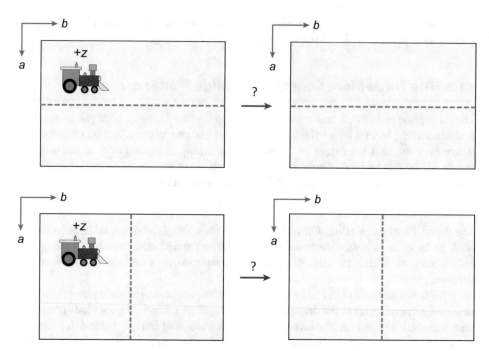

■ **Fig. 5.7** The starting configurations of the locomotive for which a glide plane *c* is to be applied, in which the mirror component should be either the (*b*,*c*) (top) or (*a*,*c*) (bottom) plane. Where is the locomotive transferred to? The solution can be found in Appendix A.1.2

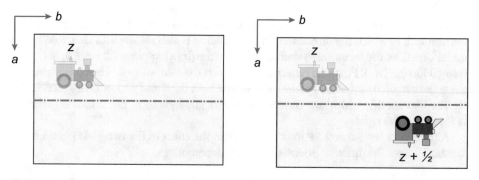

■ **Fig. 5.8** Impact of a diagonal glide plane *n*, which in this case is oriented perpendicular to the drawing plane. The translation is carried out by +½ in the direction *b* and *c*

means that this must also be reflected, whereby it is moved to the original side of the mirror component and then must be translated again by ¼ along the (*b*,*c*) plane (locomotive ③). The glide plane *d* must be applied also to this locomotive at this location, resulting in locomotive ④, ■ Fig. 5.10, bottom right. And what would be the result if the glide plane

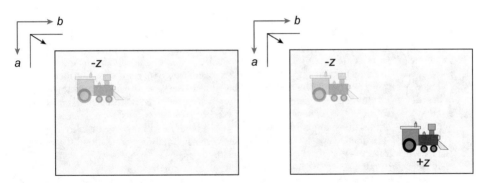

Fig. 5.9 The effect of a diagonal glide plane *n* which is in the drawing plane on a locomotive. The translation is carried out by +½ in direction *a* and *b*

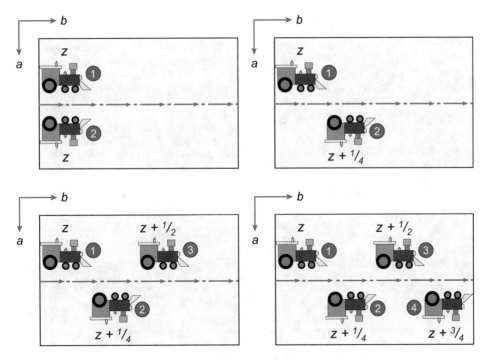

Fig. 5.10 Illustration of the impact of a diamond-like glide plane *d* (see text)

d were now also applied to the locomotive ④? Exactly – it would come to rest with respect to the (a,b) plane or x and y coordinates, respectively, on the locomotive ①, however displaced by a whole unit cell dimension in the positive z direction, which means nothing else than that it sits in the neighboring cell – in relation to the original locomotive, it emerged by pure translation in the z direction.

5

Generating Objects, Multiplicities, and General and Special Positions

Talking about "generating or multiplying objects" may require explanation, as you may ask yourself whether the objects (atoms, molecules, locomotives, motifs) do not already exist. In a given crystal structure, of course, the atoms are already present. The symmetry of a structure can be viewed from two sides. One starts with a given structure, where you now have the task of specifying the symmetry elements. For example, if two atoms/objects are localized mirror-symmetrically to each other, in the example of the ◘ Fig. 5.11a the two red triangles, then you quickly discover the corresponding mirror plane. On the other hand, it is also possible to specify a certain symmetry framework (set of symmetry elements) in a still "empty structure." If you now add a motif to this structure, you have to comply with the given symmetry elements. As a consequence, not only one motif but several motifs might be created.

If two mirror planes are present, as in the fictitious example of ◘ Fig. 5.11b, and one motif is added according to ◘ Fig. 5.11c, then three further motifs must be created according to ◘ Fig. 5.11d. The two mirror planes have quadrupled the motif, or in other words, the originally first drawn motif has a multiplicity of four. This quadruplication is the effect of the existing symmetry elements.

Generally speaking, it is sufficient to specify only a few atoms in a crystal structure with respect to their position in the unit cell, while the application of the symmetry framework then allows to derive all further positions of the atoms in the unit cell.

The multiplicity of a position or an atom at a corresponding position depends on the position within the unit cell. In the example of ◘ Fig. 5.11, the multiplicity was four. However, if an atom was lying directly on the horizontal mirror plane (◘ Fig. 5.12b), the multiplicity would be only two. And if an atom was simultaneously located on both mirror planes, i.e., directly in the center, the intersection of the two mirror planes (◘ Fig. 5.12c), the multiplicity would amount to only one, because an atom on a mirror plane is only transferred onto itself and not duplicated.

The positions that do not lie on symmetry elements are called general positions. They have the maximum multiplicity of a given symmetry framework. Special positions, on the other hand, are positions that lie on symmetry elements and may have a lower multiplicity. However, objects that are located on symmetry elements with a translation component, e.g., on a glide plane, are still being multiplied – namely, by the translation component. The terms "general/special position" and "multiplicity" will be discussed again and deepened in ▶ Chap. 6.

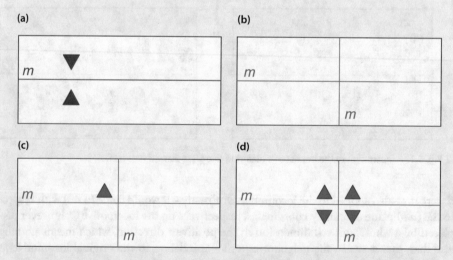

◘ **Fig. 5.11** Illustration that explains what is meant by generating or multiplying objects by a symmetry framework (see text)

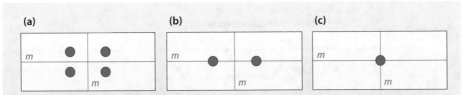

□ Fig. 5.12 Objects that lie on general positions are not located on any symmetry element; they then have the maximum multiplicity, here four (**a**). If they are located on one (**b**) or more symmetry elements (**c**), the multiplicity is reduced accordingly, here to two (**b**) or one (**c**). However, remember that this multiplicity reduction does not apply to symmetry elements that have a translational component, which also includes the glide planes

While glide planes with a translation component of ½ create one other motif from a given motif, the special glide plane d with a translation component of ¼ generates three further motifs, so that a set of four motifs are generated.

5.1.2.3 Glide Plane e

The final glide plane that has be discussed, the glide plane e, is the most complicated one. This is due to the fact that it is not a single glide plane but that it comprises two glide planes simultaneously, in which the direction of the translation is different, but the mirror components are identical or have the same orientation, respectively. Because a glide plane e is composed of two individual glide planes, it is also called a *double* glide plane.

The following example illustrates the glide mirror plane e, in which it should be composed of the simultaneous presence of a glide plane b and a glide c. In the corresponding scenario in □ Fig. 5.13, it is graphically symbolized with a dash-point-point line (a mnemonic aid: *two* points = *double* glide plane), which should be oriented perpendicular to the drawing plane (= (a,b) plane).

It starts with the locomotive ①, located below the drawing plane at height z. First, the glide plane b is applied: after the reflection, which does not change the height z, the locomotive is still below the drawing plane, and then it is translated by ½ in the direction of the b axis (locomotive ②). Now the glide plane c must be applied, both to locomotives ① and ②. The application to locomotive ① creates locomotive ③, whose height is now $z + ½$, because the glide direction runs along the c axis, and the application of the glide plane c to locomotive ② produces locomotive ④. Finally, we would have to apply the glide plane b to the new locomotives ③ and ④, both being located above the drawing plane. But as you can see in □ Fig. 5.13, bottom drawing, these two locomotives are already automatically connected by the glide plane b. This means that there are two pairs of locomotives which are related by symmetry by the glide plane b, namely, each of the two locomotives below and above the drawing plane (① + ② and ③ + ④), while the relationship by symmetry of the pairs ① + ③ and ② + ④ is given by the glide plane c.

5

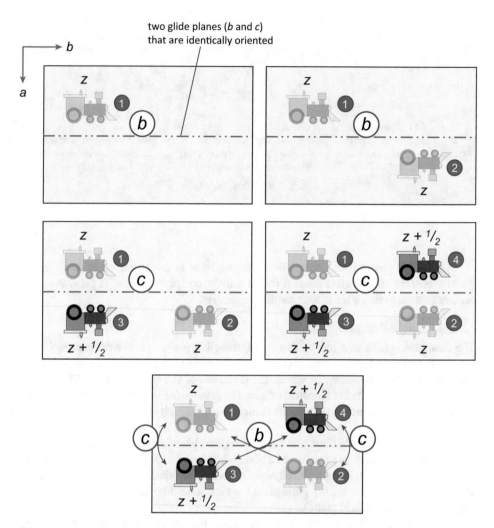

□ **Fig. 5.13**   Illustration of the impact of the double glide plane *e*, explanations in the text

With this, the glide planes are dealt with. Examples of glide planes in real crystal structures can be found in ▶ Chap. 7.

All that remains is another (translational) symmetry element that can occur (not only) in crystals: the screw axis.

5.2 **Screw Axes: Helical Symmetry**

Screw axes are – like glide planes – symmetry elements whose corresponding symmetry operation requires two operations, which are performed one after the other and necessarily together, one of which is accompanied by a translation. Thus, screw axes also belong to the symmetry elements with a translational component.

Fig. 5.14 Examples of objects with helical symmetry: spiral staircase, suspension strut (Klaus Nahr, Flickr, CC BY-SA 2.0), noodles, cooling coil of a laboratory refrigeration unit

We will see that helical symmetry is described with screw axes; therefore, they occur in objects that form a helix, where helix is only the Greek word for screw. Screws or screw axes, respectively, can be found in very different contexts, not only in crystals: spiral staircases, springs, shock absorbers, cooling coils, noodles, and much more (**■** Fig. 5.14).

A characteristic of helical objects is that their diameter is constant. This differentiates them from spiral-like objects, which have a continuously increasing or decreasing diameter (depending on the direction from which they are viewed).[1] The young shoot of a vine plant shown in **■** Fig. 5.15 has a particularly clear spiral structure.

There are also helical objects in which several helices are intertwined. The most prominent example in biology is probably the DNA (deoxyribonucleic acid, a biomolecule that carries the genetic information), which has a two-fold interweaving helical structure, i.e., forms a double helix. A beautiful sculptural model of DNA from the Principe Felipe Science Museum in Valencia is shown in **■** Fig. 5.16.

This motif was also taken up by the Danish-Icelandic artist Ólafur Elíasson, who is known for his sculptures and large format installations: in front of an office building in Munich (Ganghoferstrasse 29) stands his "endless staircase," which represents a double spiral staircase. It conveys an endless cycle of up and down and is the only staircase in the world on which there is no need to change direction when climbing up or down (**■** Fig. 5.17). The author of this book is fascinated by the works of Elíasson and would like

---

1   In this context the more correct term for a spiral staircase would be circular staircase.

Fig. 5.15   Young sprout of a vine plant.
(Dirk van der Made, Wikipedia, CC BY-SA 3.0)

to recommend a visit to his webpage or one of his exhibitions; his homepage can be reached at the following URL: ▶ www.olafureliasson.net.

5.2.1 Characteristics of Helices

Besides their diameter, helices are essentially characterized by three properties:
1. The helix or screw axis, around which the helix winds, runs in the center of a helix. Each tangent of any point along the helix forms a constant angle with the central axis, or, in other words, the gradient is constant.
2. The pitch of the helix is the length measured parallel to the axis from one turning point to the next, i.e., the height of a complete turn (■ Fig. 5.18a).
3. A helix has a sense of rotation or a handedness; it can be either left- or right-handed: if we look at the helix along the helix axis and the helix moves away from us when turning clockwise, it is a right-handed helix (and vice versa).

Left- and right-hand helices behave like image and mirror image and thus belong to the so-called chiral (gr. *cheir* = hand) objects (■ Fig. 5.18b). The handedness of a helix is an intrinsic property and does not depend on the viewing direction from which it is looked at. A right-handed helix cannot be converted into a left-handed helix unless you look at it in the mirror.

Finally, it should be noted that a helix in the strict sense of the word is something continuous, such as the handrail of a spiral staircase. Single steps of spiral stairs however are discontinuous; they only would represent a real helix, if we drew an imaginary curved line between them. Since crystals consist of discrete atoms, it is easy to see that even helix-like atomic structures in crystals do not represent a real helix. This has consequences for the rotational component of the symmetry element screw axis, as we will see in a moment.

◘ Fig. 5.16 Sculptural model of the DNA in the Principe Felipe Science Museum, Valencia, Spain

5.2.2 The Symmetry Element Screw Axis

The corresponding symmetry operation of the symmetry element screw axis is a screw rotation. It describes a coupled operation in which first a rotation around an axis and then a translation parallel to the axis are performed (◘ Fig. 5.19). If we look at a real helix, we can easily see that the rotation part can represent any angle (◘ Fig. 5.20, left). But in crystals, the rotational part is limited to certain values. That is firstly because the crystal is composed of discrete atoms (◘ Fig. 5.20, right). The second reason is, on the other hand, that the rotational symmetry in crystals is generally restricted to the order 1–4 and 6 (see ▶ Chaps. 6 and 8).

5.2.3 Screw Axes in Crystal Structures

◘ Figure 5.21 shows a first example of how screw axes can occur in crystals. It depicts a unit cell from two different views and illustrates a crystal structure of a crystal belonging to the hexagonal crystal system. We see in the left image of ◘ Fig. 5.21 that discrete helices

■ Fig. 5.17 "Endless Stairs" at KPMG Rechtsanwaltsgesellschaft in Munich, Ganghoferstraße 29. Art work by Ólafur Elíasson. Official title: Umschreibung, 2004. (Oliver Raupach, Wikipedia, CC BY-SA 2.5)

■ Fig. 5.18   (**a**) A helix is characterized by the fact that it has a constant diameter, winds around a screw axis located in the center, forms a constant angle with this, and has a certain pitch. (**b**) Left-handed and right-handed helices behave like image and mirror images

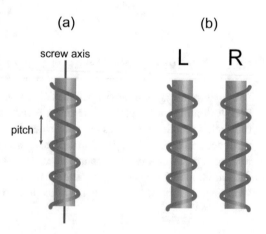

wind along the edges of the unit cell, which are running along the c axis. If we look at the (a,b) plane from above, the screws look like flat hexagons (■ Fig. 5.21, right). Because it is a crystal, the helices are present at all edges of all unit cells (arisen by the pure translation operations). But as with glide planes, there is now another symmetry element in addition to the pure translation, which has a translational component that is smaller than a com-

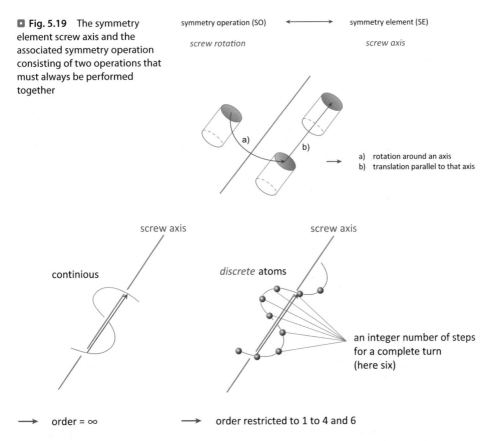

symmetry operation (SO) ⟷ symmetry element (SE)

screw rotation   *screw axis*

a)   b)

a) rotation around an axis
b) translation parallel to that axis

screw axis   screw axis

continious   *discrete* atoms

an integer number of steps
for a complete turn
(here six)

⟶ order = ∞   ⟶ order restricted to 1 to 4 and 6

■ **Fig. 5.20** In contrast to a continuous screw or helix (left), the order of rotations for screws consisting of discrete atoms is limited to the values 1–4 and 6 (right)

plete unit cell dimension: a screw axis. Along the edges, which in this case are identical to the geometric object of the screw axis, the atoms that are generated by the application of the screw can be transferred into each other: specifically in this case, any atom of the helix must first be rotated by 60° about the axis, and then it must be shifted by 1/6 along the *c* axis – since exactly six atoms form a complete turn. In this example, there is a six-fold screw axis, precisely a 6_1 screw axis, because the reciprocal of the two digits specifies the translational component: we take the index 1 as numerator and the order of the rotational component as denominator, thus obtaining 1/6.

We can generalize this (■ Fig. 5.22): a screw axis n_m, where *m* and *n* are always integers and *m* is always smaller than *n*, describes a corresponding operation in which an object must first be rotated by 360°/*n* about the axis and then has to be translated parallel to the axis by *m*/*n* units of an entire unit cell length. Furthermore, by definition, the rotation should always be carried out in the sense of a right-handed coordinate system. The rotational direction therefore results from the fact that the *x* axis is rotated toward the *y* axis. The translation is then always parallel to the *z* axis, i.e., perpendicular to the (*x*,*y*) plane.

Another example may illustrate the specification of screw axes: a 3_2 screw axis means that the order of the rotation is three, i.e., initially a rotation by 120° has to be carried out before a

5

screw axis

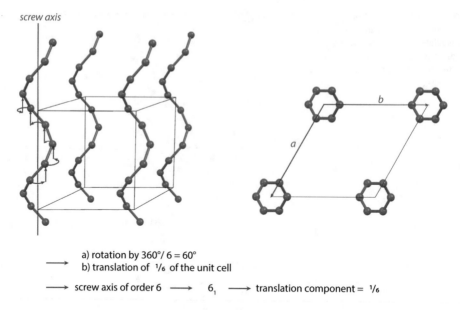

a) rotation by 360°/ 6 = 60°
b) translation of ¹⁄₆ of the unit cell

screw axis of order 6 $\longrightarrow$ 6₁ $\longrightarrow$ translation component = ¹⁄₆

◻ **Fig. 5.21** Example of a crystal structure containing a 6₁ screw axis as symmetry element. The atoms wind around the screw axis along the crystallographic c axis, where they can be transferred into each other by first rotating them 60° around the axis and then shifting them by 1/6 along the c axis (left). On the right, the top view of the (a,b) plane and the translation vectors, which transfer the individual helices as a whole into one another, are shown

◻ **Fig. 5.22** Summarizing graphic concerning the general specification of screw axes

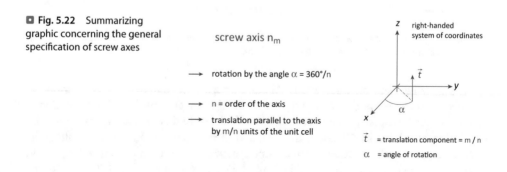

screw axis n_m

$\longrightarrow$ rotation by the angle $\alpha = 360°/n$

$\longrightarrow$ n = order of the axis

$\longrightarrow$ translation parallel to the axis by m/n units of the unit cell

right-handed system of coordinates

$\vec{t}$ = translation component = m / n

α = angle of rotation

²⁄₃ translation is performed perpendicular to it. Which screw axes can occur in crystals? Possible screw axes are 2₁, 3₁, 4₁, 4₂, 6₁, 6₂, and 6₃. In addition, there are further screw axes, which are mirror images of some of the aforementioned ones, namely, 3₂ (mirror-imaged to 3₁), 4₃ (mirror-imaged to 4₁), 6₄ (mirror-imaged to 6₂), and 6₅ (mirror-imaged to 6₁).

5.2.3.1 Tellurium: A Chemical Element with a Screw Axis

Screw axes are relatively common in molecular crystals or extended solid-state structures consisting of several elements. However, their presence in pure chemical elements is very rare. One of the few chemical elements with a screw axis is tellurium. It has a 3₁ screw axis and possesses a very interesting crystal structure. It crystallizes in the space group $P3_121$,

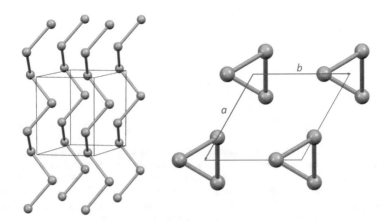

☐ **Fig. 5.23** Two views of the crystal structure of metallic tellurium. The tellurium atoms form helical chains (the tellurium-tellurium spacing is 2.835 Å), which wind along a 3_1 screw axis along the crystallographic c axis (left). On the right, a top view of the (a,b) plane is shown. The lattice parameters are $a = b = 4.456$ Å, $c = 5.291$ Å, $\alpha = \beta = 90°, \gamma = 120°$

which belongs to the trigonal crystal system.[2] In ☐ Fig. 5.23 two different views of the unit cell are shown – completely analogous to the example of the 6_1 screw axis at the beginning of this section – in the left picture, we see helices coiling around the crystallographic c axis of the hexagonal cell, while in the right picture, a projection onto the (a,b) plane is shown in which these helices look like triangles.

What is particularly interesting about the structure is the following: if we consider not only the nearest neighboring of the tellurium atoms that build this 3_1 screw axis but also the next-nearest neighboring atoms, then we recognize that it consists of hexagonal layers in the (a,b) plane, which are not simply stacked primitively, i.e., directly above each other along the c axis, but that there is a small offset from layer to layer, so that only every third next layer is identical to a given starting layer. This is exactly what manifests the presence of a 3_1 screw axis, which is highlighted in ☐ Fig. 5.24 by the tellurium atoms, which are connected with red lines.

5.2.4 Systematic Overview of Crystallographic Screw Axes

In the following a systematic overview of the screw axes that can occur in crystals will be presented, namely, in the form of point patterns that have the corresponding symmetries. In addition, the graphical symbol is given with which the screw axes can be marked. The existing screw axes with a given order of the rotational component (n) can be derived in a way that the index (m) will be varied systematically on the condition that it is an integer and smaller than the order.

---

2   You cannot yet fully decipher that space group symbol, but you already know the P and know that it must be a primitive Bravais lattice. And obviously the 3_1 screw axis is also found in the symbol. See ▶ Sect. 5.3 for more information about space groups and space group symbols.

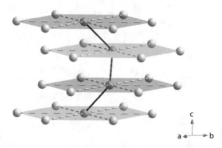

□ Fig. 5.24   If we look not only to the nearest but also to the next-nearest neighbors of a given tellurium atom, we see that they have a hexagonal environment in the (*a,b*) plane. However, these hexagonal layers do not lie directly on top of each other but are slightly shifted from layer to layer parallel to the (*a,b*) plane. Only every fourth layer is congruent. In other words, just as the tellurium atoms form 3_1 helices parallel to the *c* direction, so do the hexagonal coordination environments. One of the 3_1 screws is highlighted by blue tellurium atoms connected by red bonds

□ Fig. 5.25   A section of a point pattern having a 2_1 screw axis as a symmetry element. The associated operation consists of a rotation by 180° and a subsequent translation by ½ in the positive *c* or *z* direction

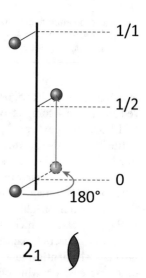

5.2.4.1   Two-Fold Screw Axes

The result of the systematic procedure is that there is only one two-fold screw axis: the 2_1 screw axis. It consists of a rotation by 360°/2 = 180° and a translation by ½ perpendicular to it (□ Fig. 5.25). The graphical symbol consists of an ellipse with two small curved hooks.

5.2.4.2   Three-Fold Screw Axes

If we now turn to the consideration of screw axes with three-fold rotational symmetry (rotation always by 120°), there are already two different screw axes: the 3_1 and 3_2 screw axes. They behave like image and mirror image to each other and thus represent a so-called enantiomorphic (from gr. *enantios* = opposite and *morph* = shape) pair. If we look at the two screw axes from the direction of the eye symbol in □ Fig. 5.26, we see the mirror-image arrangement of the point patterns.

That the two three-fold screw axes behave like image and mirror image to each other, i.e., represent helices of opposite handedness, is perhaps not immediately obvious, because

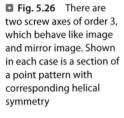

Fig. 5.26 There are two screw axes of order 3, which behave like image and mirror image. Shown in each case is a section of a point pattern with corresponding helical symmetry

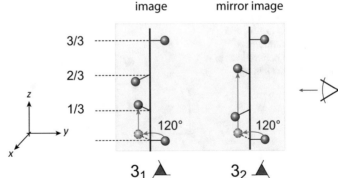

Fig. 5.27 Alternative consideration of the fact that the two screw axes of order three behave as image and mirror image. The 3_2 helix also results if we take it as an "inverse" 3_1 screw axis, i.e., if we first turn the points "forbiddenly" by $-120°$ and then shift the points of the point pattern by 1/3 in the positive z direction parallel to the screw axis

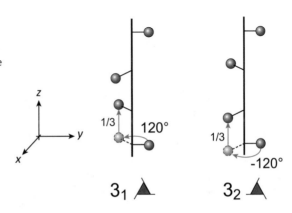

the rotation of both screw axes takes place in the same direction, namely, by definition in the sense of a right-handed coordinate system (rotation of the x onto the y axis), and the translation part is different (⅓ vs. ⅔). How can this be understood? Well, for an explanation we temporarily ignore the crystallographic rule and try to interpret the 3_2 screw axis differently: the corresponding point pattern can also be reproduced by first rotating by $-120°$ and then translation of ⅓ in the positive z direction (■ Fig. 5.27). This immediately shows that the 3_1 and 3_2 screw axes actually describe helices with opposite handedness! However, because a negative rotation value is not permitted, this must be "compensated" for via the translational part.

5.2.4.3 Four-Fold Screw Axes

In addition to the two enantiomorphic screw axes 4_1 and 4_3, which again describe helices with opposite handedness, there is one further four-fold screw axis, namely, the 4_2 screw axis (■ Fig. 5.28), in which the motif is first rotated by 90° and then translated by 2/4 = ½ in the direction of the positive c axis. Note that although it is allowed to shorten the fraction that represents the amount of the translation, it is not allowed to make a supposedly equivalent 2_1 screw out of a 4_2 screw! A 4_2 screw axis implies a rotation by 90°, a 2_1 screw axis by 180°! However, a 4_2 screw axis automatically also contains a simple two-fold axis of rotation.

Fig. 5.28 Overview of the screw axes of order 4. The two screw axes 4_1 and 4_3 form an enantiomorphic pair and behave like image and mirror image

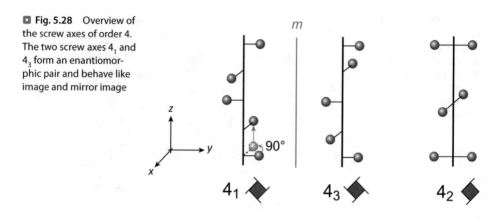

Fig. 5.29 Overview of all screw axes with rotational symmetry of order 6, each shown on a section of a point pattern. There are the two enantiomorphic pairs: 6_1 as well as 6_5 and 6_2 and 6_4. For the screw axis 6_3, there can be no mirror image

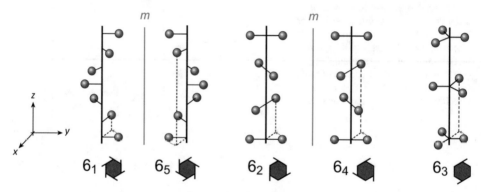

5.2.4.4 Six-Fold Screw Axes

Among the six-fold screw axes, there are the two enantiomorphic pairs (6_1 and 6_5 as well as 6_2 and 6_4) (rotation always by 60°) which each behaves like image and mirror image. In addition, there is the 6_3 screw axis, which automatically includes a simple three-fold axis of rotation. ■ Figure 5.29 shows all conceivable six-fold screw axes in the form of corresponding point patterns.

Up to this point, all conceivable symmetry elements that can occur in crystals have been treated. Thus, we have laid the foundation for the next chapter, in which the totality of the symmetry of crystalline structures is explained: the so-called space groups.

5.3 Space Groups

In this section we climb to the final stage in the systematization of crystal structures – the space groups (■ Fig. 5.30).

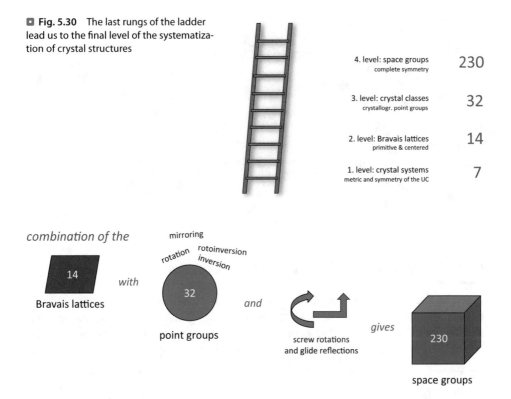

Fig. 5.30 The last rungs of the ladder lead us to the final level of the systematization of crystal structures

4. level: space groups
complete symmetry
230

3. level: crystal classes
crystallogr. point groups
32

2. level: Bravais lattices
primitive & centered
14

1. level: crystal systems
metric and symmetry of the UC
7

combination of the

14

Bravais lattices

with

mirroring

rotation   rotoinversion
   inversion

rotation

32

point groups

and

screw rotations
and glide reflections

gives

230

space groups

Fig. 5.31 How the different symmetry levels lead to the 230 space groups

On the path to this point, we had started with the 7 crystal systems and went over to the 14 Bravais lattices, which describe the 14 possibilities of point arrangements when we restrict ourselves to pure translations as a symmetry element. The examination of macroscopic crystalline objects showed that, discounting their potentially additional internal translational symmetry, 32 basically different types of symmetry exist, which are used to describe the outer shapes of crystals – the 32 crystal classes or crystallographic point groups, respectively. If we now combine all point symmetry elements – inversion, rotation, reflection, mirroring, and rotary inversion – with the pure translation symmetry of all possible lattices and the symmetry elements with translational components (glide planes and screw axes), we get exactly 230 different arrangements or types of periodic patterns in three dimensions (**Fig. 5.31**). This was shown almost simultaneously and independently by the two scientists Schoenflies [1] and Fedorov [2].

The number of possible symmetry types depends on the dimensionality of the space we are looking at (**Table 5.1**). 230 sounds like a relatively large number, and you may think that it is difficult to learn how to deal with all these space groups. You may wonder why we do not live in a two-dimensional world in which only 17 plane groups exist. On the other hand, we can be glad that we are not living in a four-, five-, or six-dimensional world, in which we would have to deal with approx. 5,000, 220,000, or 29 million space groups!

◻ **Table 5.1** Dependence of the number of crystal systems, Bravais lattices, and space groups on the dimensionality of "space"

	1	2	3	4	5	6
Crystal systems	1	4	7	40	59	251
Bravais lattices	1	5	14	74	189	841
Space groups	2	17	230	4894	222,097	28,927,915 $(+ x)^a$

[a]The number given refers to the number of space groups without consideration of the enantiomorphic pairs; the correct determination of the number including all enantiomorphic pairs is still pending

5.3.1 Definition and Notation of Space Groups

Before we look at concrete examples of space groups and take a detailed look at the "Bible of Crystallographers" – the *International Tables for Crystallography*, which contains a systematic listing of all 17 plane and 230 space groups in Volume A – the term space group should be defined. Furthermore, we will introduce the handling of the symbolism of space groups.

> The term space group can be relatively simply defined as a set of symmetry elements – together with their respective symmetry operations – which *completely* describes the spatial arrangement of a 3D periodic pattern. Nothing more!

The nomenclature of space groups is very similar to that of plane (wallpaper) groups, already discussed in ► Chap. 4. It consists of a maximum of four symbols for each space group. In the so-called short notation, it can also be less. The following example (◻ Fig. 5.32) of the space group $Pca2_1$ – which belongs to the orthorhombic crystal system – should clarify the basic structure. The space group symbol always begins with the Bravais type, i.e., the indication of the type of centering. This can be P (primitive = not centered), I for body-centered, C or A or B for the case that a one-side face centering is present, or F if an all-side face centering is given. The next three symbols specify symmetry elements with respect to three given directions, the so-called viewing directions, which vary from crystal system to crystal system. The first viewing direction in the orthorhombic system is parallel to the a axis. Here, a glide plane c is present; mirror and glide planes are always oriented perpendicular to the viewing direction, which results from the fact that the direction of a surface is indicated by its normal vector and this in turn is perpendicular to the surface. Along the second viewing direction – in the orthorhombic system, this is the b direction – there is a glide plane a. And finally, along the third viewing direction c, there is a 2_1 screw axis; screw axes and axes of rotation are always oriented parallel to the viewing direction.

Please note that all letters in a space group symbol are set italic, but all numbers are set normally.

It may come as a surprise that only symmetry elements are specified in three directions in addition to the information about the existing centering. You may ask yourself: are

Fig. 5.32 Constitution of a space group symbol and explanation of how to read it, using space group $Pca2_1$ as an example. ‖ means "parallel to," and ⊥ means "perpendicular to." All letters in a space group symbol are italicized, and all numbers are set normally

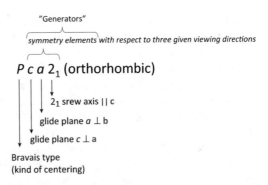

"Generators"

symmetry elements with respect to three given viewing directions

$P c a 2_1$ (orthorhombic)

2_1 srew axis ‖ c

glide plane a ⊥ b

glide plane c ⊥ a

Bravais type
(kind of centering)

Table 5.2 Overview of the specified viewing directions for the seven crystal systems

Crystal system	Viewing directions		
Triclinic	none		
Monoclinic	b $(c)^a$	–	–
Orthorhombic	a	b	c
Tetragonal	c	a	[110]
Trigonal	c	a	–
Hexagonal	c	a	[210]
Cubic	c	[111]	[110]

[a]In the monoclinic crystal system, symmetry elements can only appear in one direction; by convention this should be the b direction; however, in some countries, especially in Eastern Europe, the c direction is used instead

these three symmetry elements sufficient to completely describe the symmetry of a crystal in a given space group? Aren't there also space groups that contain 8, 16, or even 48 symmetry elements? Indeed, there are! And yet symmetry elements in the three viewing directions are sufficient, because the symmetry elements specified in this way are the so-called *generators*; together they form the generating system. And this means that not all symmetry elements have to be specified but only the set of symmetry elements from which all others can be derived. You already know for instance that two mirror planes being perpendicular to each other will automatically create a two-fold axis of rotation along their cutting edge. And there are numerous other laws, e.g., that the presence of certain symmetry elements automatically determines the presence of other symmetry elements, so that three elements of symmetry or three viewing directions are actually sufficient.

As mentioned above, the viewing directions vary from crystal system to crystal system – we simply have to learn them by heart. They have already been discussed in the crystallographic point groups in ▶ Sect. 3.3.2, but for reasons of practicability, they will be briefly repeated here (**Table 5.2**).

In the triclinic crystal system, there is no viewing direction at all, because in the triclinic crystal system, there is maximum an inversion center that has no direction.

In the monoclinic crystal system, symmetry elements – besides a center of inversion – can occur only along one direction which, by definition or according to the international standard, should be the b direction.

In the orthorhombic crystal system, everything is very simple: the first, second, and third viewing direction is along the a, b, and c direction.

In the tetragonal, trigonal, and hexagonal crystal system, the axis of rotation of highest order should determine the c direction. For all of these crystal systems, the second viewing direction is the a direction. But the third viewing direction is not the remaining third crystallographic direction b. Because the rule of thumb applies that you should look out for something "new." This means that you have to choose the viewing directions in such a manner that symmetry elements can be found that are not already given by the other viewing directions. As a reminder, for the three crystal systems mentioned, the a and b axes are identical, for reasons of symmetry! This means that if we first look along the direction a and, for example, find a mirror plane, then we must necessarily also discover a mirror plane along the b direction – but this is not new information about existing symmetry elements. For this reason, a viewing direction must be selected that reveals new symmetry elements. These directions are given as lattice vectors in ◻ Table 5.2 (see also ▶ Sect. 3.3.2).

In the cubic crystal system, the first viewing direction is along the c axis, the second viewing direction is along the space diagonal, and the third direction is running along the face diagonal.

5.3.2 Relationship of Space Group, Point Group, and Crystal System

Before we look at a first concrete example of a space group in detail, the relationship between the space group, the resulting crystallographic point group and the associated crystal system will be discussed briefly.

The first step is to understand how the crystallographic point group can be derived from the space group. There is a relatively simple recipe for this, which consists of only two or three points:

— First of all, it should be noted that the point group describes the symmetry of macroscopic, finite objects; since such objects lack any translational symmetry, the Bravais type can be omitted in the first step. Centering simply does not occur because there is no lattice.

— Since all other symmetry elements with a translational component describe internal symmetry, too, but should be ignored in the case of macroscopic objects, the translation-dependent symmetry elements must therefore be converted into their corresponding symmetry elements without translational component:

    — Glide planes become simple mirror planes.
    — Screw axes are converted to simple axes of rotation.

— All symmetry elements that do not have a translational part are simply retained, because they already describe point symmetry. Therefore, axes of rotation, rotoinversion axes, and mirror planes remain unchanged.

Let's practice this recipe with three examples:

1. A crystal should belong to space group $P2_1/n$. What is the crystal class? The P – for primitive – is omitted, and the 2_1 screw axis is converted into a simple two-fold axis

of rotation and the diagonal glide plane into a normal mirror plane. This results in the point group of $2/m$.

2. A substance may crystallize in the space group *Fddd*. For the crystal class, the type of centering is disregarded, and the three diamond-like glide planes become simple mirror planes: the point group is therefore *mmm*.

3. The space group used to explain the basic structure of the space group symbolism was *Pca2₁*. Applying the above recipe, the point group results in *mm2*.

Apart from the fact that we are now able to easily derive the point group from the space group, the question arises: why should we do this? Is that an end in itself? No, it can make sense to make these derivations for two reasons: first, the point group immediately tells us something about the morphology of a crystal crystallizing in a given space group! Although we are not able to derive from the space group how a concrete crystal, a specific find of a mineral, looks exactly, but with the information of its point group, we know the maximum symmetry of its outer shape! Mind you the maximum symmetry, but sometimes it can also be much less symmetrical. Secondly, even if this is of minor importance, in the crystallographer's everyday life, it can be useful to derive the point group from a given space group in order to quickly obtain clarity about the crystal system to which the space group belongs. It is much easier to internalize the affiliation of the 32 crystal classes to the 7 crystal systems than to that of the 230 space groups. The crystal class *mmm* is quickly determined from the space group *Fddd*, which – with a certain amount of experience – makes it clear that *Fddd* belongs to the orthorhombic crystal system and not to the cubic one.

Based on a few variations of lattices and motifs, the connections between the three levels of consideration, namely, space group, point group, and crystal system, should now be illustrated.

Example 1

1. *Variant*

Suppose that there is a crystal lattice or unit cell, as shown in ◻ Fig. 5.33. The following applies to the lattice parameters: $a = b$ and $\alpha = \beta = \gamma = 90°$.

First, we can see that lattice points are only at the corners of the unit cell, i.e., it is a primitive unit cell, and the Bravais type is *P*. The metric of the lattice is indicative of a tetragonal cell. Next, we look for symmetry elements and discover along the *c* direction a four-fold axis of rotation, confirming the tetragonal crystal system. Perpendicular to the same direction,

◻ **Fig. 5.33** A unit cell with the specified metric, in which lattice points are only at the corners and for which the space group and crystal class should be specified

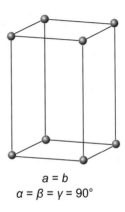

$a = b$
$\alpha = \beta = \gamma = 90°$

◘ Fig. 5.34   The unit cell from ◘ Fig. 5.33 with superimposed symmetry elements

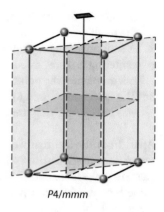

P4/mmm

◘ Fig. 5.35   (a) Variation of the unit cell from ◘ Fig. 5.33 with respect to the motif, which now consists of tetragonal pyramids. (b) Superposition of the unit cell with the symmetry elements

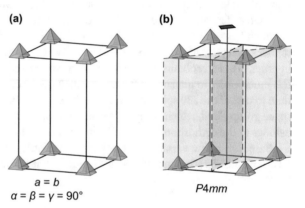

(a)

$a = b$
$α = β = γ = 90°$

(b)

P4mm

another mirror plane can be found. The further viewing directions in the tetragonal crystal system are along the *a* direction – perpendicular to this direction another mirror plane is present – and along the face diagonal [110], and here, too, we encounter a mirror plane perpendicular to it. All symmetry elements found are superimposed in ◘ Fig. 5.34 in the unit cell. Thus, the space group is clearly determined to be $P\frac{4}{m}mm$ (the plain text notation is P4/mmm). According to the algorithm, this results in the crystal class 4/mmm.

2. Variant
In this variant, the metric should remain the same, but instead of the spheres or atoms, the motifs now consist of tetragonal pyramids (◘ Fig. 5.35a). What has changed? Well, the fourfold axis of rotation along the *c* axis is also present here, but the mirror plane perpendicular to it is missing! The mirror planes along the *a* direction and along [110] are still present (◘ Fig. 5.35b). This results in the space group *P4mm* and the crystal class *4mm*.

3. Variant
The motif should be unchanged with regard to the second variant, but there is an additional pyramid present in the center of the unit cell. All pyramids should have the same orientation (◘ Fig. 5.36a). Clearly, now a body centering is present. Nothing changes with regard to the presence of all other symmetry elements (◘ Fig. 5.36b), so the space group is now *I4mm*. The crystallographic point group does not change; it remains *4mm*.

◘ Fig. 5.36 **(a)** Another variation of the unit cell from ◘ Fig. 5.33; the metric is unchanged like the motif, but now there is another motif in the center of the cell. **(b)** The symmetry elements are the same as in the cell in ◘ Fig. 5.35, but the Bravais type is different

(a)

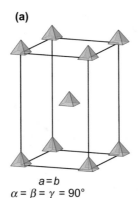

$a = b$
$\alpha = \beta = \gamma = 90°$

(b)

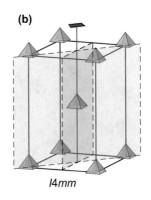

I4mm

◘ Fig. 5.37 **(a)** Another variation of the unit cell from ◘ Fig. 5.33 in which the motif is mirrored in the center. **(b)** The symmetry elements are unchanged with respect to the unit cell in ◘ Fig. 5.36, but the Bravais type is only primitive. Note that the (overall) motif has changed

(a)

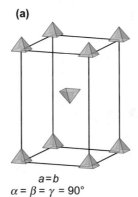

$a = b$
$\alpha = \beta = \gamma = 90°$

(b)

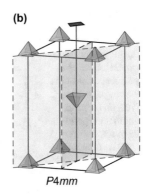

P4mm

4. *Variant*

In this variant, one of the tetragonal pyramids is inverted, namely, that in the center of the unit cell (◘ Fig. 5.37a). What are the consequences? There is no longer a body centering present because for this all motifs at all locations must have the same orientation. At first, this might be somehow counterintuitive to some readers, but centering is basically nothing else than translations. The only action that is allowed, when going from a lattice point at a corner to a lattice point on a face or to one in the center is: carrying out a translation. However, here also the orientation of the motif, which represents the lattice point, is changed. Therefore, the motif at one of the corners cannot be congruently mapped onto the motif in the center. So there is actually a primitive cell present, because the motif has changed (it now consists of the inverted pyramid in the center and eight eighths of the pyramids at the corners). Since all pyramids are nevertheless oriented in such a way that they are compatible with the four-fold rotational symmetry, this is retained. The same applies to the two mirror planes along the *a* and [110] direction (◘ Fig. 5.37b). This results in the space group *P4mm* and the crystal class *4mm*.

◻ Fig. 5.38 A primitive unit cell with the specified metric whose space group and crystal class should be determined

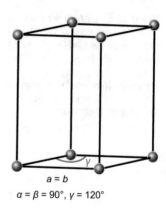

$a = b$
$\alpha = \beta = 90°, \gamma = 120°$

◻ Fig. 5.39 **(a)** There is a mirror plane (in green) perpendicular to the six-fold axis of rotation; **(b)** two further mirror planes can be found, one perpendicular to the viewing direction [100] (in red) and another perpendicular to the direction [210] (in orange)

(a)

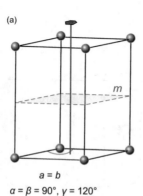

$a = b$
$\alpha = \beta = 90°, \gamma = 120°$

(b)

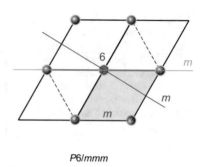

P6/mmm

Example 2

In another example, the relationship between the symmetry of the lattice, which abstracts from the motif, and the symmetry including a motif should be clarified. The hexagonal crystal family, which combines the tri- and hexagonal crystal system, both of which have the same metric ($a = b$ and $\alpha = \beta = 90°$ and $\gamma = 120°$), is suitable for this purpose. In ▶ Sect. 2.7 and 2.8 we worked out the difference between the two crystal systems. The characteristic feature of the trigonal crystal system is the three-fold rotational symmetry, that of the hexagonal, the six-fold rotational symmetry.

1. *Variant*

Let us first consider a primitive lattice with the metric given above (◻ Fig. 5.38) – which symmetry elements are you able to discover along the three crystallographic viewing directions and what is the space group?

Since we see only one unit cell, discovering the symmetry elements is less easy than in the previous example (see also ▶ Sect. 3.3.2). Obviously, there is a mirror plane perpendicular to the *c* axis (◻ Fig. 5.39a, green); and we know that there is a six-fold axis of rotation running along the lattice vector *c*. The other symmetry elements can be visualized more

□ **Fig. 5.40** Variation of the motif compared to the unit cell from
□ Fig. 5.38, it now consists of trigonal prisms

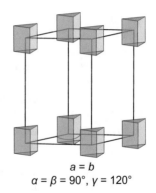

$a = b$
$\alpha = \beta = 90°, \gamma = 120°$

□ **Fig. 5.41** (a) The mirror plane perpendicular to the *c* axis is still before. (b) Superimposed symmetry elements in the 2D projection of the unit cell; instead of a six-fold axis of rotation, only a three-fold one is now present here; furthermore, compared to the unit cell of
□ Fig. 5.39 one of the mirror planes disappeared

(a)

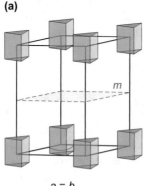

$a = b$
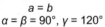
$\alpha = \beta = 90°, \gamma = 120°$

(b)

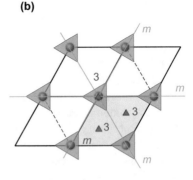

P31m

easily if a 2D projection is used and parts of the neighboring cells are also displayed: there is one mirror plane perpendicular to the *a* direction (red) and another one perpendicular to the direction [210] (orange in □ Fig. 5.39b). So, the space group is *P6/mmm* – the space group of the hexagonal crystal system with the highest symmetry. The crystal class can be determined to 6/*mmm*; the long notation would be 6/*m* 2/*m* 2/*m*.

2. *Variant*
But we haven't chosen a (real) motif yet! Let us assume that the motif consists of trigonal prisms with their long axis parallel to the *c* axis (□ Fig. 5.40) – the author does not think of any chemical structure that could be described as such, but this is irrelevant for our considerations! Instead of a six-fold axis of rotation, we now only have a three-fold axis of rotation. Perhaps not immediately obvious to everyone is the fact that the three-fold axes are not only running along the vertical edges (in 2D projection in □ Fig. 5.41b, they are shown at the corners of the cell), but there are also two three-fold axes present inside the cell. However, let us, first of all, record the following important message:

> The motif is not compatible with a six-fold axis of rotation, but with the metric of the cell, which corresponds to that of the hexagonal system! In other words, the lattice always shows the maximum symmetry of a crystal system or a crystal family. Which crystal system it belongs to within a family, or if it shows the maximum symmetry within a crystal system, is dependent on the motif represented by the lattice point. In this context, you should also try to convince yourself that each lattice – regardless of the actual crystal system – must necessarily have a center of inversion, of course, the crystal structure does not. Now you also fully understand ☐ Table 1.1 in ▶ Chap. 1!

The other existing symmetry elements in our example are a mirror plane, which is perpendicular to the c direction (green), but which does not belong to the generating system, and finally another unique mirror plane, namely, the mirror plane perpendicular to the direction [210] (orange); for a better overview, the two other mirror planes generated by the three-fold axis of rotation are also shown in ☐ Fig. 5.41. The space group is $P31m$ and the crystal class $3m$.

This last section of this chapter should give you a sense of how the interplay of lattice and centering on the one hand and the self-symmetry of the motifs on the other – beginning with individual atoms that have, as spheres, any symmetry and ending with objects of lower symmetry – results in different combination possibilities of an overall symmetry. The different space groups are exactly the description for that. However, we have also seen that it is by no means trivial to derive all existing symmetry elements from the space group symbol, since only the generating symmetry elements are specified. And you know only roughly where these symmetry elements are located in the unit cell, if you know the crystallographic viewing directions of the associated crystal system. You shouldn't invest too much time to derive the position and orientation of all other symmetry elements, because you can conveniently take this information from the "Bible of Crystallographers," the *International Tables for Crystallography*, especially Volume A [3], in which all space groups with their specifications are listed systematically. So, it's high time to familiarize yourself with this most important tool of the crystallographers in their everyday work.

References

1. Schoenflies A (1891) Krystallsysteme und Krystallstructur. B.G. Teubner, Leizpig. https://www.archive.org/details/krystallsysteme00schogoog. Accessed 29 Sept 2019
2. Fedorov ES (1971) Symmetry of crystals. (1891) Translated from Russian by David und Katherine Harker, New York, American crystallographic association monograph no. 7. American Crystallographic Association, New York
3. Hahn T (ed) (2005) International tables for crystallography, Volume A, space-group symmetry, 5th corr. Ed. Springer, Dordrecht

Space Groups II: International Tables

© Springer Nature Switzerland AG 2020
F. Hoffmann, *Introduction to Crystallography*, https://doi.org/10.1007/978-3-030-35110-6_6

In this chapter, you will become familiar with the handling of the International Tables for Crystallography, Volume A (ITA for short) [1]. Since this is a fairly abstract and dry matter, it is best to read this chapter only if you encounter an interesting structure of a certain space group and decide that you are really interested in looking up all the details of this space group in the International Tables.

Volume A of the International Tables ( Fig. 6.1) has the simple title "Space-group symmetry." In this volume, in addition to the 17 plane groups, all 230 space groups with all symmetry elements and further details are systematically listed. As a crystallographer, I almost daily look up certain things in these Tables. It is a very useful compendium, but unfortunately it is also very expensive. Fortunately, there is an Internet resource that also contains all essential symmetry information about the space groups, which you can access if you do not have access to the International Tables (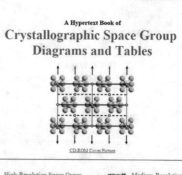 Fig. 6.2 shows a screenshot of the start page) [2].

Volume A of the International Tables contains, among other things, the following information about space groups:

— The space group symbol in long as well as in short notation and a unique number (from 1 to 230)
— The crystallographic point group derived from the space group, both in the notation according to Schoenflies and according to Hermann and Mauguin

Fig. 6.1   The reference work par excellence and daily tool for crystallographers: The famous International Tables for Crystallography, here Volume A with the title: "Space-group symmetry." (Reproduced with kind permission of the © International Union of Crystallography,
▶ http://it.iucr.org/)

Fig. 6.2   Screenshot of the homepage of an electronic and free alternative to the International Tables, A. (© Birkbeck College, University of London)

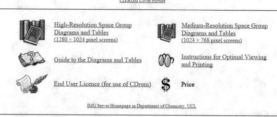

- The crystal system to which the space group belongs
- A diagram in which all symmetry elements are drawn into a unit cell
- A diagram of the general (atomic) positions
- Information about the origin of the cell
- A list of all symmetry operations as coordinate transformations or all principally different positions (point positions) in the cell, respectively, including the information about the multiplicity, the Wyckoff symbol, and the site symmetry

In addition, some more data can be found that belong either to the field of X-ray diffraction, i.e., the structural analysis of crystals using X-rays, or to advanced topics in the field of group theory (e.g., super/subgroup relationships). With the exception of a brief explanation of the group term, these two areas are not dealt with further in this book.

6.1 The Space Group *Pmm*2 in Detail

As a first example, let us look at the entries for the space group *Pmm*2. It is the space group no. 25. This space group was chosen because the degree of complexity is still relatively low and the information on the symmetry and construction of the diagrams is easy to understand. Having laid this foundation, we will then discuss more complicated examples.

First, in ◘ Fig. 6.3 a downscaled overview of the two pages reserved for space group *Pmm*2 is shown. These will now be discussed section by section, although – for didactic reasons – we will not strictly follow the order of the information presented.

6.1.1 The Header Section of a Space Group

In the header section (◘ Fig. 6.4) of each space group, the long and short symbols are given, which are identical in the case of space group *Pmm*2, as well as a unique number, here 25. All space groups are systematically numbered from 1 to 230. This should also guarantee a clear identification in those cases where careless mistakes were made in the spelling of the symbol.

In the middle of the header, the crystal class or crystallographic point group is indicated, both in Schoenflies (C_{2v}) and in Hermann-Mauguin notation (*mm*2). The superscript one at the Schoenflies symbol has nothing to do with labeling of symmetry, but merely states that it is the first space group of this crystallographic point group. On the right side of the header, the crystal system to which the space group belongs is given, as well as the Patterson symmetry, which belongs to the area of X-ray structure analysis, the background of which, however, will not be discussed here. Purely formally, the Patterson symmetry is derived from the space group symmetry by (1) initially preserving the Bravais type, (2) adding an inversion center to the overall symmetry, but (3) omitting the translational component of all further symmetry elements; for instance, a 2_1 screw will become a simple two-fold axis of rotation.

6.1.2 Area of Diagrams/Diagram of the Symmetry Elements

In this area, four diagrams are depicted (◘ Fig. 6.5), three of which (top left, top right, bottom left) show the symmetry elements in a schematically drawn empty unit cell. These three diagrams differ only in the choice of the projection plane. The most important of these three

6

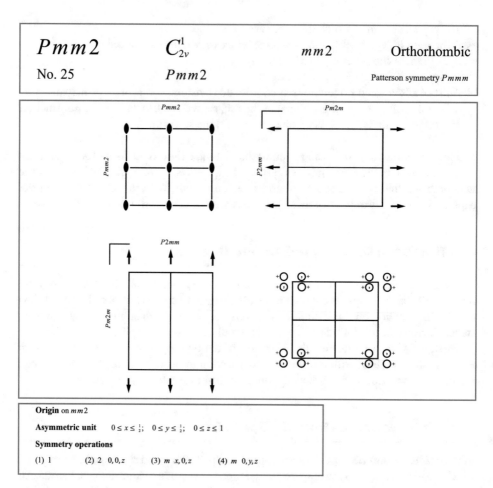

Pmm2 C_{2v}^1 $mm2$ Orthorhombic

No. 25   Pmm2   Patterson symmetry $Pmmm$

Origin on $mm2$

Asymmetric unit $0 \leq x \leq \frac{1}{2};\ \ 0 \leq y \leq \frac{1}{2};\ \ 0 \leq z \leq 1$

Symmetry operations

(1) 1   (2) 2 0,0,z   (3) m x,0,z   (4) m 0,y,z

▫ **Fig. 6.3** Overview of the two pages for the space group *Pmm*2 in the ITA. The areas which are discussed in detail are highlighted by blue frames. (Reproduced with kind permission of the © International Union of Crystallography, ▶ http://it.iucr.org/)

diagrams is the one shown at the top left. It presents the symmetry element diagram in the standard projection, which is shown enlarged again in ▫ Fig. 6.6. This diagram shows the projection of an orthorhombic unit cell onto the (a,b) plane with all symmetry elements at their respective positions within the cell. The bold lines are mirror planes (▫ Fig. 6.6a). There are two horizontal mirror planes that run along the edges of the unit cell and are oriented parallel to the (b,c) plane. These are actually identical, because they result from a translation of a complete unit cell length along the a direction. Furthermore, there is a horizontal mirror plane that runs through the middle of the cell. Incidentally, their existence is due to the presence of the two horizontal mirror planes at the edges of the cell. In addition to the three horizontal planes, there are also three vertical mirror planes that run parallel to the (a,c) plane (▫ Fig. 6.6a). And therefore – according to the principle of symmetry that two mirror planes perpendicular to each other generate a two-fold axis of rotation along their edge of

CONTINUED   No. 25   *Pmm*2

Generators selected   (1); $t(1,0,0)$; $t(0,1,0)$; $t(0,0,1)$; (2); (3)

Positions

Multiplicity, Wyckoff letter, Site symmetry		Coordinates			Reflection conditions
					General:
4	i	1	(1) x,y,z (2) $\bar{x},\bar{y},z$ (3) $x,\bar{y},z$ (4) $\bar{x},y,z$		no conditions
					Special: no extra conditions
2	h	m..	$\frac{1}{2},y,z$ $\frac{1}{2},\bar{y},z$		
2	g	m..	$0,y,z$ $0,\bar{y},z$		
2	f	.m.	$x,\frac{1}{2},z$ $\bar{x},\frac{1}{2},z$		
2	e	.m.	$x,0,z$ $\bar{x},0,z$		
1	d	mm2	$\frac{1}{2},\frac{1}{2},z$		
1	c	mm2	$\frac{1}{2},0,z$		
1	b	mm2	$0,\frac{1}{2},z$		
1	a	mm2	$0,0,z$		

Symmetry of special projections

Along [001] $p2mm$ Along [100] $p1m1$ Along [010] $p11m$
$\mathbf{a}' = \mathbf{a}$ $\mathbf{b}' = \mathbf{b}$ $\mathbf{a}' = \mathbf{b}$ $\mathbf{b}' = \mathbf{c}$ $\mathbf{a}' = \mathbf{c}$ $\mathbf{b}' = \mathbf{a}$
Origin at $0,0,z$ Origin at $x,0,0$ Origin at $0,y,0$

Maximal non-isomorphic subgroups

I   [2] $P1m1$ (Pm, 6)   1; 3
       [2] $Pm11$ (Pm, 6)   1; 4
       [2] $P112$ ($P2$, 3)   1; 2

IIa   none

IIb   [2] $Pma2$ ($\mathbf{a}' = 2\mathbf{a}$) (28); [2] $Pbm2$ ($\mathbf{b}' = 2\mathbf{b}$) ($Pma2$, 28); [2] $Pcc2$ ($\mathbf{c}' = 2\mathbf{c}$) (27); [2] $Pmc2_1$ ($\mathbf{c}' = 2\mathbf{c}$) (26);
       [2] $Pcm2_1$ ($\mathbf{c}' = 2\mathbf{c}$) ($Pmc2_1$, 26); [2] $Aem2$ ($\mathbf{b}' = 2\mathbf{b}, \mathbf{c}' = 2\mathbf{c}$) (39); [2] $Amm2$ ($\mathbf{b}' = 2\mathbf{b}, \mathbf{c}' = 2\mathbf{c}$) (38);
       [2] $Bme2$ ($\mathbf{a}' = 2\mathbf{a}, \mathbf{c}' = 2\mathbf{c}$) ($Aem2$, 39); [2] $Bmm2$ ($\mathbf{a}' = 2\mathbf{a}, \mathbf{c}' = 2\mathbf{c}$) ($Amm2$, 38); [2] $Cmm2$ ($\mathbf{a}' = 2\mathbf{a}, \mathbf{b}' = 2\mathbf{b}$) (35);
       [2] $Fmm2$ ($\mathbf{a}' = 2\mathbf{a}, \mathbf{b}' = 2\mathbf{b}, \mathbf{c}' = 2\mathbf{c}$) (42)

Maximal isomorphic subgroups of lowest index

IIc   [2] $Pmm2$ ($\mathbf{a}' = 2\mathbf{a}$ or $\mathbf{b}' = 2\mathbf{b}$) (25); [2] $Pmm2$ ($\mathbf{c}' = 2\mathbf{c}$) (25)

Minimal non-isomorphic supergroups

I   [2] $Pmmm$ (47); [2] $Pmma$ (51); [2] $Pmmn$ (59); [2] $P4mm$ (99); [2] $P4_2mc$ (105); [2] $P\bar{4}m2$ (115)

II   [2] $Cmm2$ (35); [2] $Amm2$ (38); [2] $Bmm2$ ($Amm2$, 38); [2] $Imm2$ (44)

■ **Fig. 6.3** (continued)

---

*Pmm*2 C_{2v}^1 $mm2$ Orthorhombic

No. 25   *Pmm*2   Patterson symmetry *Pmmm*

■ **Fig. 6.4** Header section for the space group *Pmm*2 in the ITA. (Reproduced with kind permission of © International Union of Crystallography, ▶ http://it.iucr.org/)

---

intersection – there must also be a two-fold axis of rotation at each intersection edge, which is oriented perpendicular to the plane of projection (■ Fig. 6.6b).

The additional symmetry element diagrams at the top right and bottom left of the area of diagrams (■ Fig. 6.5) are two further projections, namely, on the (a,c) plane (top right)

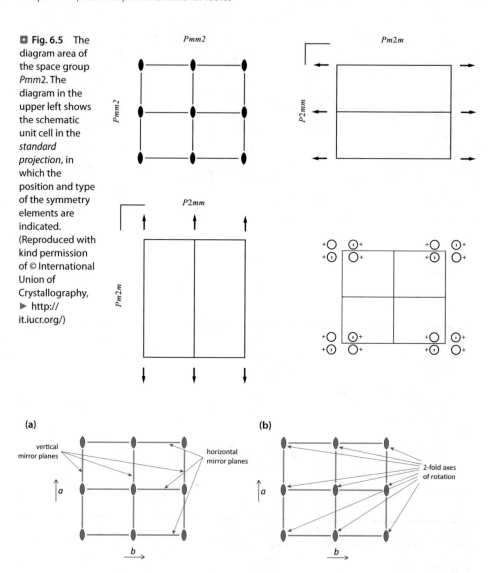

■ **Fig. 6.5** The diagram area of the space group *Pmm2*. The diagram in the upper left shows the schematic unit cell in the *standard projection*, in which the position and type of the symmetry elements are indicated. (Reproduced with kind permission of © International Union of Crystallography, ▶ http://it.iucr.org/)

■ **Fig. 6.6** The symmetry element diagram for the space group *Pmm2* in the standard projection: (a) mirror planes, (b) two-fold axes of rotation

and the (*b,c*) plane (bottom left).[1] You already know the two angles, which are shown in the upper left corner, from the consideration of the glide planes; here they are graphical symbols for simple mirror planes which lie in the same plane as the projection plane. The two-fold axes of rotation, which no longer run perpendicular to but in the projection plane, are marked as arrows.

---

1   It remains incomprehensible why these diagrams in the International Tables have not been labeled with axes in the conventional sense (*a*, *b*, *c*). The derivation at which plane you are currently looking at must be derived through the space group symbolism next to the diagrams, where the order of the viewing directions has been changed: for example, the direction indicated by *P2mm* corresponds to the *c* direction. The author hopes that this deficiency will be corrected in the next edition.

6.1.3 The General Position Diagram

The general position diagram is always located in the lower right corner of the area of diagrams (◩ Fig. 6.5). In ◩ Fig. 6.7 it is shown again in enlarged form. This diagram is also drawn in the standard projection; here we again look at the (*a,b*) plane. In the general position diagram, a circle represents a motif of the crystal structure, for example, an atom or an atom group. If you have a crystal structure in front of you in which all atoms are on general positions and you align the structure according to the standard projection direction, here along the *c* direction, then you should see a pattern that essentially corresponds to the pattern of the general position diagram. Please note that your cell axis ratios can be different and that your atom or atom group positions do not have to exactly match the relative positions of the circles shown here.

The general position diagram may look a bit confusing at first glance. In fact, it is relatively easy to construct. Therefore, in the following, we want to create this diagram step by step by starting with a single position and then successively reconstructing the diagram by applying the symmetry framework.

This means that we superimpose the symmetry elements on the general position diagram, and we apply the corresponding symmetry operations of the space group until no more circles are added. The only condition is that we actually start with a *general position*. Remember: a general position is a position within the cell which is not located on a symmetry element.

We will start with the position shown in ◩ Fig. 6.8a. The plus sign means that the circle should be above the projection plane. A second position is created by applying the red mirror plane (◩ Fig. 6.8b). If the central horizontal mirror plane is now applied to both positions, the configuration shown in ◩ Fig. 6.8c is created. Interestingly, no further circles will be added within the cell, no matter which symmetry operation we would use. For example, the application of the central two-fold axis of rotation (◩ Fig. 6.8d) would convert the circle at the top right into the one at the bottom left and the circle at the top left into the bottom right (and vice versa, of course). Alternatively, we could have used the two-fold axis of rotation axis after applying the vertical mirror plane to generate all positions in the cell; then the horizontal mirror plane would only transfer existing positions into each other.

◩ **Fig. 6.7** The diagram of the general positions of the space group *Pmm*2

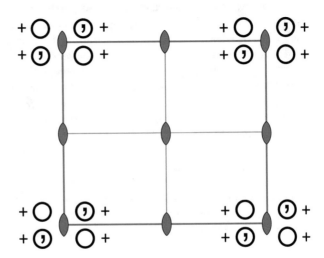

Fig. 6.8 First steps of the construction of the general position diagram by starting with only one position (a) and first considering the mirror planes (b), (c). After applying the mirror planes, all positions within the cell are already created – the application of the two-fold axis of rotation in the center of the unit cell only transfers sites that already exist (d)

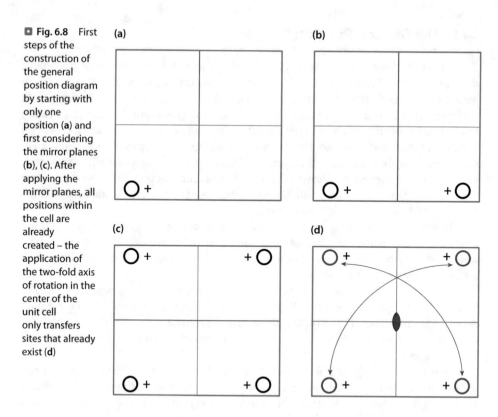

The general position diagram does not only show the general positions within the unit cell. Rather, it is only complete when all symmetry elements have been applied, including those at the edge of the cell. No sooner said than done: the applications of the two-fold axes of rotation at the corners of the cell to the circles closest to the corners are shown in the sequence in ◻ Fig. 6.9. Their application to the circles not nearest to the corner would generate copies far away from the originally considered cell and can be safely ignored.

Up to now, the two-fold axes of rotation lying on the edge bisectors have not been taken into account. Their exemplary application on two circles is shown in ◻ Fig. 6.10a, resulting in two new circles or positions. The result of applying the central two-fold axis of rotation to these newly created circles can be seen in ◻ Fig. 6.10b. If these processes are repeated for the other circles, the picture is completed. To complete the configuration from ◻ Fig. 6.10a, we could also have applied the mirror planes.

The position of the circles and how the symmetry elements affect them are thus explained. Remember: the plus signs indicate that the circles along the c axis should be somewhere above the projection plane (a,b). And since all circles carry these plus signs, this means that their height along the c axis or their fractional z coordinate does not change due to any symmetry operation.

Still an explanation is lacking, why there are some circles with a comma inside. These commas come into play by the fact that the circles may not necessarily represent individual, spherical atoms, but can consist also of asymmetric or chiral motifs, i.e., those objects which possess a non-congruent mirror image. Each application of a mirror plane transforms an object into its mirror image. This pair – image and mirror image – is graphically symbolized

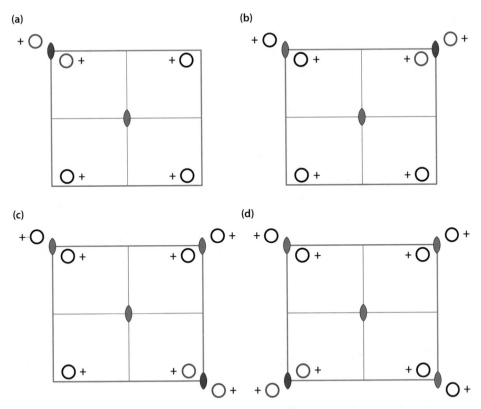

■ **Fig. 6.9** Sequence for further completion of the general position diagram in which the two-fold axes of rotation at the corners of the unit cell are successively applied to the nearest circles (highlighted in red)

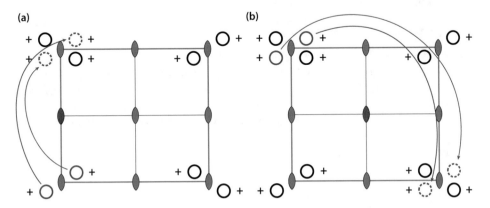

■ **Fig. 6.10** The general position diagram is completed by the application of the other two-fold axes of rotation

in the way that one of the circles does carry a comma inside and the mirrored circle does not. This means, mirroring of an empty circle leads to a circle with a comma; mirroring of a circle with a comma leads to one without. To illustrate these interrelationships, we can use a well-known motif: our locomotive (■ Fig. 6.11). Each mirror plane transfers the locomotive

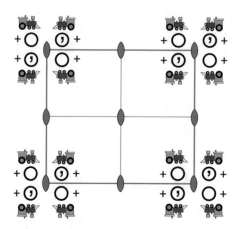

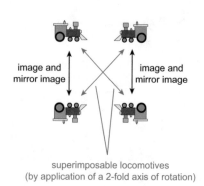

image and
mirror image

image and
mirror image

superimposable locomotives
(by application of a 2-fold axis of rotation)

◻ Fig. 6.11   Illustration of the meaning of commas in the general position diagram: they express that a reflection of a chiral motif creates a mirror image that cannot be superposed onto the original image (see text)

in its corresponding mirror image, which is non-congruent with the original image. But remember that the successive application of a vertical and horizontal mirror plane reverses the configuration twice so that the second copy is identical with the initial motif and can be superimposed by application by the two-fold axis of rotation with the initial locomotive.

6.1.4 **The Asymmetric Unit**

Another entry defines the so-called asymmetric unit in the International Tables. For the space group $Pmm2$, we find the following entry:

Asymmetric unit $0 \leq x \leq \frac{1}{2}$; $0 \leq y \leq \frac{1}{2}$; $0 \leq z \leq 1$;

This entry defines a specific part (of the volume) of the unit cell as an asymmetric unit. For the triclinic, monoclinic, and orthorhombic crystal system, the asymmetric unit – like all unit cells – forms a parallelepiped, except that it does not cover the entire unit cell, but only a certain part. In the case of the trigonal, hexagonal, and, in particular, cubic crystal system, however, the shape of the asymmetric unit can be considerably more complicated.

The asymmetric unit can be defined as follows:

> The asymmetric unit is the part of the unit cell that is sufficient to reproduce the contents of the whole unit cell by applying all symmetry operations of the space group.

It is, thus, the smallest building block of a crystal structure, from which the entire crystal can be derived. We have learned that the whole crystal can be built up by adjoining unit cells in all three spatial directions or, to be precise, along the three crystallographic direc-

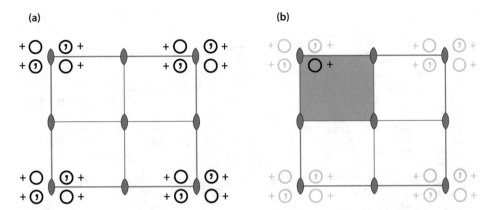

◻ Fig. 6.12 (a) The boundaries of the asymmetric unit are indicated by thin lines within the unit cell, here highlighted in red; (b) a quarter of the unit cell represents the asymmetric unit in the space group *Pmm*2. One of the four possible quarters is highlighted in red. In this quarter, exactly *one* of the four circles of the whole unit cell can be found corresponding to one quarter of the multiplicity of a general position of four

tions. But beyond the pure translations from one lattice point to another, the other symmetry elements of the space group also act on the positions/motifs *inside* the unit cell, whereby they are multiplied. We have seen how we could construct the general position diagram precisely by doing this: by specifying a position and then gradually applying the symmetry elements to this first position to derive all other positions. Here, in this example, one position is sufficient to derive all other positions. This corresponds to exactly one quarter of the unit cell.

This quarter, or four possible quarters, is also plotted in the general position diagram, by the thin lines highlighted in red in ◻ Fig. 6.12a. These lines do not represent mirror planes in this diagram, but the boundaries of the asymmetric unit. The numerical form of the asymmetric unit expresses that the range for the values of x and y is between 0 and ½ and the values of z are between 0 and 1. Thus, the upper left part of the unit cell is defined as an asymmetric unit (see ◻ Fig. 6.12b). In other words, it does not matter where that first circle is actually placed in the diagram, as long as it stays in that quarter. But all other quarters are also valid boundaries of the asymmetric unit of the space group *Pmm*2 and completely interchangeable.

Both from the definition and the wording a general property of the asymmetric unit immediately follows: it must not have internal symmetry, in any case, no "crystallographically effective" symmetry. From this, some further properties can be derived regarding the position of symmetry elements on the one hand and the boundaries of the asymmetric unit on the other:

— Mirror planes and axes of rotation must form boundary faces or edges of the asymmetric unit. Unlike mirror planes and axes of rotation, these restrictions do not apply to glide planes and screw axes, as they additionally have a translational component.

— Two-fold axes of rotation can pass through the center of a boundary face.

— Centers of inversion must either form the corners of the asymmetric unit or have to be located at the midpoints of boundary faces or edges.

Crystallographic Brain Teaser

Are you able to locate the asymmetric unit or fundamental region of a chessboard? Which of the areas highlighted in 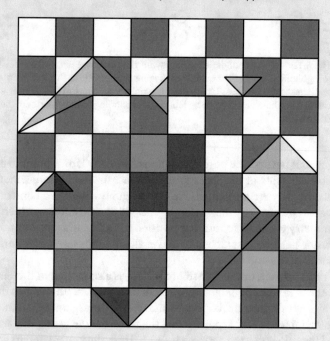 Fig. 6.13 mark asymmetric units?

To anticipate, this is a pretty challenging task. Therefore, the author would like to give you some guidance: it always makes sense to first determine the unit cell. Do not be fooled by the horizontal and vertical rows of the chessboards – the unit cell could have an orientation that does not agree with it … Then try to find all the symmetry elements of the cell and thus also determine the plane group. What is the maximum order of any existing axes of rotation? Draw the symmetry elements into the cell! Finally, remember which symmetry elements must be boundaries of asymmetric units. Now you should come to the solution that you can look up in Appendix A.1.3.

◘ Fig. 6.13 In this chessboard, different areas are highlighted in different colors. Some of them are fundamental areas, which reproduce the entire chessboard by using all symmetry elements. Which ones belong to them?

So far, the asymmetric unit has been treated either as an empty volume element or as a volume element in which there is a position at some point, in other words, as a geometric entity. In mathematical contexts, the asymmetric unit is also called the fundamental region. In crystal structures, of course, the volume element is completely filled with atoms or molecules. When it comes to molecular crystals, usually just one molecule forms the asymmetric unit, much less frequently it contains two independent molecules. However, if the molecule has intrinsic symmetry, it is also possible that the asymmetric unit contains only half a molecule or one-quarter of a molecule – the more detailed relationships are explained in ▶ Sect. 7.1.

6.1.5 Positions and Coordinates

In the section of the International Tables, which is captioned with *Positions* (◘ Fig. 6.14), the positions in the unit cell are specified again. In addition to the general positions, also the special positions are listed – and this time not in graphical form as in the general position diagram, but in the form of their coordinates.

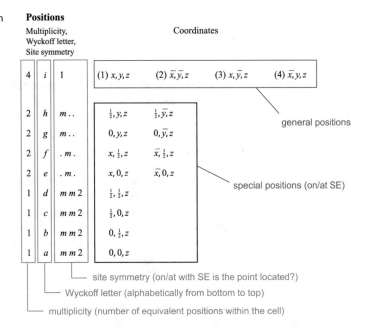

■ **Fig. 6.14** The section under which the general and special positions of the space group *Pmm*2 are specified. Explanations in the text. (Reproduced with kind permission of © International Union of Crystallography, ▶ http://it.iucr.org/)

These positions are also called Wyckoff positions or point locations, because each point P is transformed by the application of the symmetry framework into a symmetry-equivalent point P'. Of course, it depends on where the starting point is located, how many other points, and at which locations they are generated. This is expressed by the multiplicity. If the point is at a general position, there is a maximum number of equivalent points in the unit cell (maximum multiplicity). If it is at a specific position, meaning that it lies on one or more point symmetry elements (axis of rotation, rotary inversion axis, mirror plane, inversion center), then the multiplicity decreases accordingly. For a point that lies directly on a point symmetry element, that is, a symmetry element without a translational component, it is transformed into itself. Conversely, it is clear that for points that are located on symmetry elements with a translational component, the multiplicity is not decreased, because the translational component automatically ensures that a new point P' is generated, which is not congruent with P.

The *site symmetry* can therefore be interpreted in two ways:
- It indicates on which symmetry elements the position is located.
- It indicates which point symmetry this position has (or all points that are symmetry-equivalent to this position).

Therefore, the maximum point symmetry of a position can be at most that of the crystallographic point group of the space group (the most special of all special positions, so to speak). The lowest symmetry, namely, none at all, is given by the general position.

The first systematic elaboration of the point positions in the individual space groups was made by Ralph W. G. Wyckoff in 1922 [3], which is why the positions are also called Wyckoff positions. Wyckoff has labeled these positions with small Latin italicized letters (Wyckoff letters or Wyckoff symbols), starting with a small "*a*" for the most special position and then alphabetically ascending for the positions with lower symmetry. The order is, however, partly arbitrary. This is due to the fact that a point position may occur multiple times, that is, may have the same site symmetry. Therefore, in these cases, you have to look up in the International Tables which site symmetry is exactly meant. In practice, for the specification of the position of an atom, often not only the Wyckoff letter but also the multiplicity is indicated, for instance, "4a."

Some entries from the section *Positions* (◘ Fig. 6.14) will be explained in the following. In the first row, the general positions are listed in the form of their coordinates, whereby a bar above the letter is again to be interpreted as a minus. They are:

(1) x, y, z (2) $\bar{x}, \bar{y}, z$ (3) $x, \bar{y}, z$ (4) $\bar{x}, y, z$

These coordinates can be simultaneously interpreted as a coordinate transformation if the respective symmetry operations are performed on an atom/motif/object having the general position x, y, z. The first entry leaves the coordinates unchanged, that is, x, y, z is transformed into x, y, z. This is, of course, nothing else than the effect of an identity operation! The second entry converts x, y, z into $-x, -y, z$. To identify the respective symmetry operation, we can consider: which symmetry operation reverses two of the signs and leaves the third unchanged? Correct, a rotation by 180° around the origin. And since the coordinate z remains unchanged, it must therefore be a two-fold axis of rotation parallel to the c axis!

The third and fourth transformation only change the sign of one coordinate, while the other two remain unchanged. This is the effect of a mirror plane, namely, for (3) a mirror plane lying in the (a,c) plane, therefore the y coordinate is inverted, and for (4) one that is in (b,c) plane, causing the inversion of the x coordinate.

In summary, from an object placed on any general location x, y, z, the application of the symmetry framework will generate four copies, namely, at the coordinates that are listed in the first row. This is exactly the multiplicity given in the first column. And now it should also be clear what *Site Symmetry* means: it indicates with which symmetry this given position is linked. Since the general location is specified in the first row, it is clear that the object is located at a site somewhere in the cell where *no* symmetry element is located, which is expressed in crystallography with the identity, i.e., 1.

After specifying the general positions, the specifications for the special positions follow in the other rows. For the first entry, the multiplicity is given as two and the site symmetry is specified with m.. The two points are placeholders for possible symmetry elements in the second and third viewing directions in the orthorhombic crystal system, which are here, however, empty. The first viewing direction is running along the a axis; hence, the site symmetry indicates that it is a position directly on the mirror plane perpendicular to the a direction. The first coordinate specified for this particular location is ½, y, z, meaning that the fractional coordinate x is set to 0.5, but y and z are not restricted and can be initially chosen arbitrarily. So, this location is somewhere on the central horizontal mirror plane. For example, we can set $y = 0.3$ (z is unspecified). Now we apply to this circle at this position all possible symmetry elements, and the result is the second specified coordinate ½, $-y, z$. For our example, this results in the new coordinates ½, $-0.3, z$. Apparently, the new location is in the neighboring cell, in the direction of $-b$. If we want to consider a symmetry-equivalent location in the same cell, we have to add +1 for y, which yields ½, +0.7, z. ◘ Figure 6.15 illustrates the respective relationships. Now we realize that the second position is generated, for instance, by a reflection at the central vertical mirror plane or by rotation at the central two-fold axis of rotation. Because it does not matter which additional symmetry operation is carried out, no further positions will be generated, and this means that the multiplicity for this particular position is only two.

◻ Fig. 6.15 Illustration of site symmetry specified with the Wyckoff symbol 2 *h* of the space group *Pmm*2 (see text)

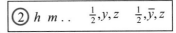

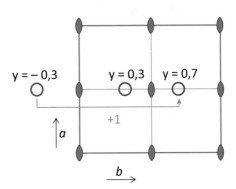

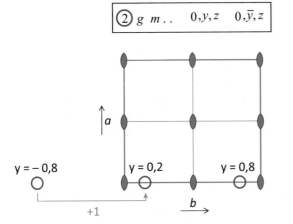

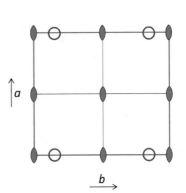

◻ Fig. 6.16 Illustration of the site symmetry 2 *g* of the space group *Pmm*2 (see text)

Similarly, in the third row, we find another special position, which is also indicated by *m* .. The first coordinate tells us that it must be a position on the top or bottom horizontal mirror plane (in fact, these two are identical, since they can be converted into each other by a translation of +1): 0, *y*, *z*. Here, too, we find the second coordinate by inverting the *y* coordinate: 0, −*y*, *z*. An example with *y* = +0.8 is shown in ◻ Fig. 6.16, left.

If we count all circles *within* the cell, however, we only get 1/2 + 1/2 = 1, while the multiplicity is given by two. What has not yet been considered is the application of the central horizontal mirror plane, the result of which is shown in ◻ Fig. 6.16, on the right. It is also possible to interpret the two additional circles that have emerged from the original circles by pure translation of +1 along the *a* direction; in fact, the two upper ones are identical to the lower two, now, also the required multiplicity of two results.

The next two lines in the table of positions are those that lie on the vertical mirror planes. Analogous to the horizontal mirror planes, two positions are indicated, one on the central and one on the mirror plane at *x* = 0. The corresponding site symmetries are consequently given as . *m* ., that is, they lie on the mirror planes that run perpendicular to the second viewing direction in the orthorhombic crystal system.

■ **Fig. 6.17** The Wyckoff position d in the space group $Pmm2$ has the multiplicity one and is located exactly in the center of the cell, where the two mirror planes intersect. In addition, the two-fold axis of rotation is running perpendicular to them

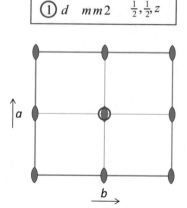

Only one kind of special position is left, namely, the position that is simultaneously located on three symmetry elements: $mm2$. This is the point of intersection of two mirror planes on which simultaneously the two-fold axis of rotation is located. The Wyckoff position d describes a position in the center of the cell or, because the height is not specified along c, a point located somewhere on the axis of rotation, but lying at ½ with respect to a and b (■ Fig. 6.17). No matter which symmetry element is applied, no further circle will be generated, and the multiplicity is exactly one.

The last three lines (Wyckoff positions $a–c$) describe, analogously, sites on the edge bisector along the a direction (Wyckoff position c), on the edge bisector along the b direction (Wyckoff position b), and finally one position which lies along a line passing through the origin (Wyckoff position a at $0, 0, z$).

6.1.6 Origin and Origin Choice

Like so much in crystallography, the choice of the origin of the unit cell in relation to the existing symmetry elements is subject to certain standards. The first rule is: the origin should either be placed on a center of symmetry (inversion center) or on a point of high site symmetry, i.e., a point on which several symmetry elements lie. For space groups which have no special position, this rule cannot be applied. Of course, this is true only for the space group $P1$, which has no symmetry at all; here, the origin can be chosen arbitrarily. However, it concerns also, for instance, the space groups of the crystal classes 2, 3, 4, and 6 as well as the space groups of the monoclinic crystal system of the crystal class m, because here, apart from a simple axis of rotation or a simple mirror plane, no further symmetry elements are present. Even at the points where several symmetry elements meet, there does not necessarily have to be a common *point* of intersection, and possibly only common faces of intersection or edges of intersection are created. Accordingly, the origin can be placed "free-floating" on a face or along a line.

The origin choice is explicitly specified in a separate entry in the International Tables. For the space group $Pmm2$, the entry is:

Origin on   *m m* 2

Since the space group *Pmm*2 has no center of inversion, the position of highest symmetry is chosen here. At this position, two mirror planes intersect, and along their cutting edge, the automatically generated two-fold axis of rotation runs, which also passes through the origin. Important here is the little word "on," which indicates that the origin is not at a *fixed point*, in which case the word "at" would have been chosen. Here, too, the intersecting mirror planes and the two-fold axis of rotation only form a common *line*.

In addition to being able to read the choice of origin directly under the "Origin" entry, there are two other ways to assure yourself of how the origin was chosen and which symmetry elements pass through it:

- It can be easily obtained by inspecting the visual diagram of the symmetry elements.
- We search under the "Positions" for the coordinate 0, 0, 0 or for a coordinate triplet, which can principally contain these three values.

For the space group present here, the most special position is the Wyckoff position 1*a* with the coordinates 0, 0, *z*. As coordinate *z* is not specified, it may also have the value 0. This in turn means nothing else than that the origin is fixed with respect to *x* and *y*, but can have any values of *z*. Now we also may look up the site symmetry of this Wyckoff position: it is *mm*2 – voila, this is consistent with the entry in "Origin!"

For those space groups, which have both a center of inversion and another position of high symmetry not identical with the position of the inversion center, the two principles for setting the origin compete with each other. In this case, the International Tables give two descriptions of the same space group, which differ exactly by the choice of the origin and thus the relative position of the symmetry elements to the origin. As a further consequence, the coordinates of the general and special positions change too. In such cases, the header section contains an additional entry which indicates what variant is presented: ORIGIN CHOICE 1 or ORIGIN CHOICE 2. In addition, the displacement with respect to the inversion center is specified under the "Origin" entry. For example, for the space group *Pnnn*, the choice of the origin is "222" (i.e., the intersection of three mutually perpendicular two-fold axes of rotation), and this point is displaced from the inversion center $\bar{1}$ by ¼, ¼, ¼:

Origin at 222, at ¼, ¼, ¼ from $\bar{1}$

The remaining sections of the space group entries of the International Tables will be discussed based on the space group *P*2₁/*c*.

6.2   **The Space Group *P*2₁/*c* in Detail**

As a second example, the space group *P*2₁/*c* should be discussed. There is a good reason for this, because it is very likely that you will encounter this space group, once you are concerned with the handling of space groups. It is by far the most frequently occurring space group; see the following background information.

Frequency of Space Groups

The Cambridge Crystallographic Data Center (CCDC) operates one of the most important crystallographic databases in which more than 1,000,000 crystal structures are now recorded – the so-called Cambridge file. Every year it publishes a statistic, which indicates how the structures are distributed among the space groups. The results for the ten most frequent representatives (as of 2018) are shown in ◻ Table 6.1. Nearly 35% of all structures have a crystal structure with the space group $P2_1/c$.

It is also very interesting to note that the ten most common space groups already account for 87.6% of all structures, and the remaining 220 space groups account for just 12.4% of all structures. The distribution over the 230 space groups is therefore extremely uneven. According to these statistics, the least represented space group is the space group $P4_2mc$ (No. 105): just two crystal structures belong to this space group. The complete statistics can be accessed via this URL: ▶ https://www.ccdc.cam.ac.uk/support-and-resources/ccdcresources/84ded717fe9c48d6a93b41701a5e7e84.pdf

There are also statistical evaluations of the corresponding frequencies of space groups in inorganic substances. A 2009 study evaluating the largest crystalline structure database for inorganic compounds (the Inorganic Crystal Structure Database, ICSD) determined the order given in ◻ Table 6.2 for the first ten ranks [4]. A comparison between the organic and inorganic crystal structures with respect to their distribution of the space groups gives interesting findings:

– Although $P2_1/c$ and $P\bar{1}$ are also among the most common space groups for inorganic substances, they account for a significantly smaller percentage of the total number of all substances.

◻ **Table 6.1** Overview of the ten most frequent space groups in the Cambridge Structural Database of the Cambridge Crystallographic Data Center (CCDC), in which 754,897 organic or organometallic crystalline substances were recorded up to January 2018

Rank	Space group no.	Space group	Number of substances	Percentage of the substances
1	14	$P2_1/c$	319,411	34.4
2	2	$P\bar{1}$	229,917	24.8
3	15	$C2/c$	77,441	8.4
4	19	$P2_12_12_1$	66,030	7.1
5	4	$P2_1$	47,603	5.1
6	61	$Pbca$	30,586	3.3
7	33	$Pna2_1$	12,778	1.4
8	62	$Pnma$	9,775	1.1
9	9	Cc	9,680	1.0
10	1	$P1$	8,905	1.0

☐ **Table 6.2** Overview of the ten most common space groups of the Inorganic Crystal Structure Database (ICSD) maintained by the Fachinformationszentrum Karlsruhe (FIZ). The evaluation is based on data from 2006, when the database comprised 100,444 structures

Rank	Space group no.	Space group	Number of substances	Percentage of the substances
1	62	$Pnma$	7,432	7.4
2	14	$P2_1/c$	7,233	7.2
3	225	$Fm\bar{3}m$	5,624	5.6
4	227	$Fd\bar{3}m$	5,123	5.1
5	2	$P\bar{1}$	4,018	4.0
6	139	$I4/mmm$	4,017	4.0
7	15	$C2/c$	3,817	3.8
8	194	$P6_3/mmc$	3,415	3.4
9	12	$C2/m$	3,412	3.4
10	221	$Pm\bar{3}m$	3,013	3.0

- Overall, the frequency of the inorganic compounds is distributed much more homogeneously over the space groups.
- For organic compounds, there are no cubic space groups among the first ten ranks, while there are three in the field of inorganic compounds. Likewise, in the case of the organic crystal compounds, there is no tetragonal or hexagonal space group among the first ten ranks, while in the case of inorganic compounds, one can find at least one entry for each of these crystal systems.

6.2.1 Settings

In comparison to the first example, the space group $Pmm2$, we find some differences in the header section for the space group $P2_1/c$ (☐ Fig. 6.18).

$$P2_1/c \qquad C_{2h}^5 \qquad 2/m \qquad \text{Monoclinic}$$

No. 14 $P12_1/c1$ Patterson symmetry $P12/m1$

UNIQUE AXIS b, CELL CHOICE 1

☐ **Fig. 6.18** The header section for space group $P2_1/c$ in the ITA. (Reproduced with kind permission of © International Union of Crystallography, ▶ http://it.iucr.org/)

The first difference is that there is an additional entry that gives information about the so-called setting of the cell: UNIQUE AXIS b, CELL CHOICE 1. What does that mean? Well, we know that the space group $P2_1/c$ belongs to the monoclinic crystal system and that symmetry elements can occur only in one viewing direction. This is the so-called unique axis, which, according to the international standard, should be the crystallographic b direction. This is also expressed in the long symbol of the space group, which is different from the short symbol and is $P12_1/c1$. In this notation, it is explicitly stated that there are no symmetry elements along the a direction (first position after the centering symbol P) and c direction (third position after the centering symbol). In the b direction, there are two symmetry elements, namely, a 2_1 screw axis parallel to the b direction and a glide plane c perpendicular to that direction. Although there is this convention of choosing the b axis as unique axis, this is by no means obligatory. This explains the existence of *variants* of the space group $P2_1/c$, which arise from the fact that the a or c axis can also be chosen as unique axis. Correspondingly, the long symbols would be $P2_1/c11$ or $P112_1/c$, and the (mostly) oblique angle would then be the angle between the b and c axis (α) or between the a and b axis (γ). While the choice of the a axis as unique axis is extremely rare, the c axis can quite often be found as unique axis, especially among authors from Eastern Europe.

But these are not the only variations with regard to the setting of the cell. There are other variants that are not related to the choice of the unique axis, but to the translation direction of the glide plane with respect to the cell axes. In the standard setting, it is the c direction, which is why it is $P2_1/c$! However, possible variants are equally $P2_1/a$ and $P2_1/n$. How can this be understood? Why can the glide direction of a glide plane be different and yet represent the same? This is because the crystal or the motifs, which are connected to each other via the glide plane, do not know anything about the crystallographic axes – these are defined by the crystallographer. And depending on the setting, the glide direction may be different with respect to the axis system. Conversely, this means that these variants must be transformable by redefining the axes. This will be shown below.

In ◻ Fig. 6.19a a schematically monoclinic cell in the standard setting is shown, in which additionally a glide plane is drawn (in gray) as well as a locomotive, the motif already known to us. In ◻ Fig. 6.19b, c, the effect of the symmetry operation – that of the glide reflection c – is shown in two steps: first the reflection and second the translation by ½ in the c direction. The translation direction is indicated also as a red arrow.

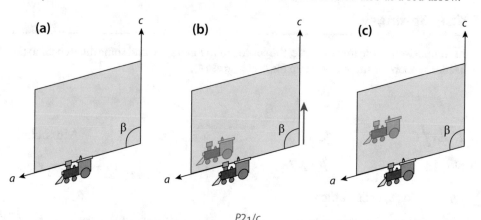

$P2_1/c$

◻ **Fig. 6.19** Effect of a glide plane c on the motif of a locomotive: **(a)** initial situation, **(b)** reflection, **(c)** translation in positive c direction

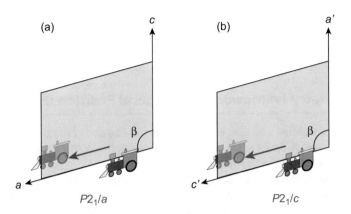

Fig. 6.20 The redefinition of the axes a and c transforms the space group $P2_1/a$ into the space group $P2_1/c$

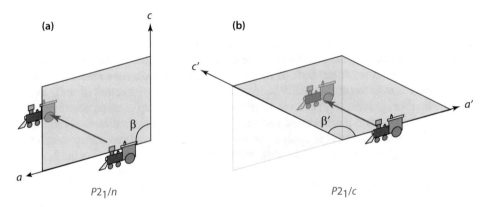

Fig. 6.21 The space group $P2_1/n$ can also be transformed into the standard space group $P2_1/c$ by an axis transformation

Now consider an analogous scenario in which the plane of the glide plane should be oriented so that the direction of translation is in the a direction (◘ Fig. 6.20a). How does the redefinition of the axes have to be done in order to generate a glide plane c? Obviously, the old a axis must become the new c axis and vice versa the old c axis the new a axis (◘ Fig. 6.20b). The oblique angle β does not change!

Now, only the variant $P2_1/n$ remains. In this case, the translational direction of the glide plane is running along the cell diagonal (◘ Fig. 6.21a). In this case, how can we redefine the axes that the diagonal becomes the new c direction, in order to get to the standard setting? By a linear combination of the previous lattice vectors $\vec{a}$ and $\vec{c}$, i.e., $\vec{c}' = \vec{a} + \vec{c}$. In addition, if the standard is to be considered that the monoclinic angle should be larger and not smaller than 90°, the direction of the old a direction must be inverted; i.e., $\vec{a}' = -\vec{a}$ (◘ Fig. 6.21b).

So, there are a total of nine(!) possible variants of the space group $P2_1/c$, six of which are common in practice, three with the unique axis c and three with the unique b axis: $P112_1/c$, $P112_1/a$, $P112_1/n$, $P12_1/c1$, $P12_1/a1$, and $P12_1/n1$.

So, if you come across a space group that seems completely strange to you, one that does not appear in the standard list of the 230 space groups, do not despair – it is certainly a nonstandard variant of a known space group.

6.2.2 Symmetry Framework and General Position Diagram

For purposes of training and deepening, we also want to derive the general position diagram for the space group $P2_1/c$ (◻ Fig. 6.22) by constructing it step by step using the symmetry framework shown in ◻ Fig. 6.23.

Let's start with the position in the upper left corner *within* the cell (◻ Fig. 6.24a), which is above the projection plane and therefore marked with a plus sign. First, let the inversion center, which lies at the origin, act on the position. The result is shown in ◻ Fig. 6.24b. Since we have performed a point *mirroring*, (a) the position is now below the projection plane (minus sign), and (b) the configuration of a possibly chiral motif is reversed, which is expressed by the comma. If we next apply the central inversion center, two new positions are generated in the lower right area of the cell (◻ Fig. 6.24c), which themselves are of course also symmetry-related to each other by an inversion center (at position 1, 0, 1). The two "position pairs" in the upper right and lower left area can be generated by the inversion centers located on the edge bisectors along the c axis (at positions 0, 0, ½ and 1, 0, ½, ◻ Fig. 6.24d). As a result, inversion centers have now also automatically been created on the edge bisectors of the a axis.

Any further application of centers of inversion will only map existing positions onto itself, and the position pairs around the edge bisectors along the c direction cannot be created. Apparently, we have to apply a previously non-considered symmetry element in order to complete the general position diagram. For example, we can choose the 2_1 screw axes, highlighted in red in ◻ Fig. 6.25. Thus, the positions are generated, which are denoted by ½ + and ½ –, which are the short notation for ½ + y and ½ – y, because the

◻ **Fig. 6.22**   The general position diagram of the space group $P2_1/c$

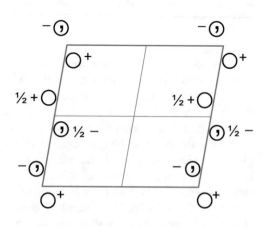

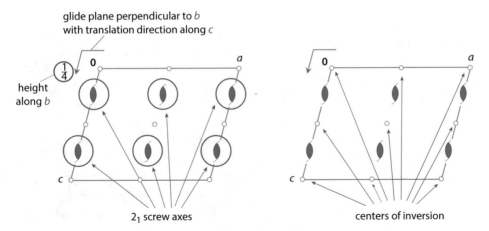

Fig. 6.23 The symmetry element diagram for the space group $P2_1/c$. Parallel to the projection plane is a glide plane, which is indicated by the oblique arrow in the upper left corner. The height along the b axis is 1/4. In fact, there is a second glide plane, which is at the height of $b = 3/4$ (There would be really enough space to include this information directly at the arrow. Instead of doing so, the International Tables prefer to hide the information that 1/4 always means 1/4 and 3/4 in a small, barely traceable note in the accompanying text to the Tables. This is a mystery to the author). There are also a number of 2_1 screw axes parallel to the b direction, two of which are *inside* the cell and two more on the edges of the c axis, at $c = 1/4$ and 3/4. Furthermore, there are inversion centers on all corners and all bisectors of the edges as well as in the center of the cell

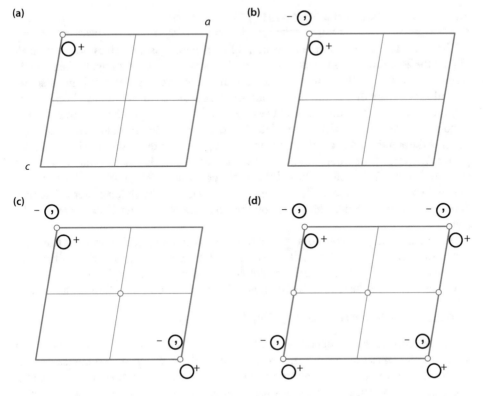

Fig. 6.24 First part of the construction of the general position diagram of the space group $P2_1/c$ using the framework of the symmetry elements (see text)

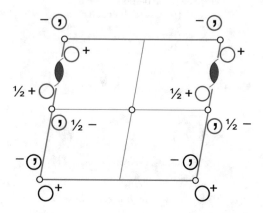

◻ Fig. 6.25 The positions marked with ½ + or ½ – can be generated by applying the 2_1 screw axis highlighted in red to the already realized sites

2_1 screw axis along the b direction leads to an increase of the fractional coordinate y by ½ due to its translational component. This completes the general position diagram – and this without applying a single glide mirroring. Is it possible that the glide plane is not present? But, of course it is – are you able to figure out which of the positions will be transferred into each other by the glide plane c? Note that according to the diagram of the symmetry elements, the glide planes are at a height of ¼ and ¾ along the b direction.

6.2.3 Generators and Symmetry Operations

We have just found out that the derivation of all general positions does not necessarily require the application of all of the symmetry elements of the space group. In the previous section, we first applied the inversion centers and were able to generate all the positions around the corners of the unit cell. Incidentally, an alternative would have been to first consider only the inversion center at the origin and then to apply the pure translational vectors of the lattice. In order to complete the picture, however, it would have required in any case the application of another symmetry element, either the 2_1 screw axis or the glide plane c. Conversely, we would not necessarily have needed the inversion center to generate all equivalent general positions. In addition to the pure translations, the additional application of the glide plane c and the 2_1 screw axes is sufficient – the inversion center would then have been created automatically again by the existence of the other two symmetry elements.

Under the entry "Generators selected," the International Tables now list the symmetry elements or operations that are sufficient for a complete generation of the general positions (and their translational-equivalent points in the neighboring cells) and in which order they have to be applied. For the space group $P2_1/c$, it looks like this:

Generators selected (1); $t(1,0,0)$; $t(0,1,0)$; $t(0,0,1)$; (2); (3)

The symmetry elements or operations are here provided with numbers in brackets, except for the whole translations from lattice point to lattice point, which are written (somewhat cumbersome) as t operations running along the crystallographic directions a, b, and c. The symmetry elements or operations with numbers are described in a separate section "Symmetry operations." For $P2_1/c$ we find:

Symmetry operations

(1) 1   (2) 2 (0, ½, 0) 0, y, ¼   (3) $\bar{1}$ 0,0,0   (4) c x, ¼, z

The first symmetry element (1) is the identity. Of course, it has to be applied first; alternatively, one could say that this is the creation of the first general position. The second symmetry element (2) is the 2_1 screw axis parallel to the b direction, with the translational part in brackets, ½ in the direction b, i.e., addition of ½ to the y coordinate. Right after the brackets, we find the specification, where this symmetry element is located (on a = 0, c = ¼, and parallel to b, that is, on all y coordinates). The third symmetry element (3) is the center of inversion at the origin. The last symmetry element (4) is the glide plane c. Here, for the sake of simplicity, the translational part is not specified because it is already in the name of the symmetry element. Again, however, the position of the plane is specified, lying in the (a,c) plane, perpendicular to b, and along the b direction at the height y = ¼.

6.2.4 Positions and Coordinates

In fact, you already know how to interpret the coordinates, multiplicities, Wyckoff symbols, and site symmetries specified in the "Positions" section. The first line specifies the general positions and looks like this:

4 e 1   (1) x, y, z (2) $\bar{x}, y+\frac{1}{2}, \bar{z}+\frac{1}{2}$ (3) $\bar{x}, \bar{y}, \bar{z}$ (4) $x, \bar{y}+\frac{1}{2}, z+\frac{1}{2}$

As already explained in the example for $Pmm2$ in ▶ Sect. 6.1, they can also be understood as coordination transformations, whereby each coordination transformation represents a corresponding symmetry element or operation. (1) and (3) can easily be recognized as identity and inversion. The 2_1 screw axis parallel to the b axis – represented by the transformation (2) – first inverts the two coordinates x and z by rotation of 180°, while the translational part increases the y coordinate by ½. However, a comparison with the transformation under (2) shows that ½ should also be added for z or -z. The reason for this is not immediately obvious, but we have to consider that the 2_1 screw axis does not run through the origin, but is shifted in the direction of c by ¼. This is exactly what is taken into account by adding +½. Analogous considerations can be made for the transformation (4), which represents the glide plane c. The reflection at a plane in the (a,c) plane inverts the y coordinate, and the translational part of the glide plane c increases the z coordinate by ½. The additional addition of ½ to the y term, however, is in turn based on the fact that the glide plane is not located at the height y = 0, but at y = ¼.

Try to understand the additional summation term ½ for the screw axis and the glide plane by making corresponding sketches. Try different values for z in the case of the screw axis or y in the case of the glide plane of a general position and check whether the specified formula delivers the correct result (this should be the case).

6.3 Overview of the Graphical Symbols of the Symmetry Elements of the International Tables

You have already been introduced to many of the graphical symbols for symmetry elements used in the International Tables. However, since we have worked out many issues within the framework of two-dimensional projections, some symbols have been used for

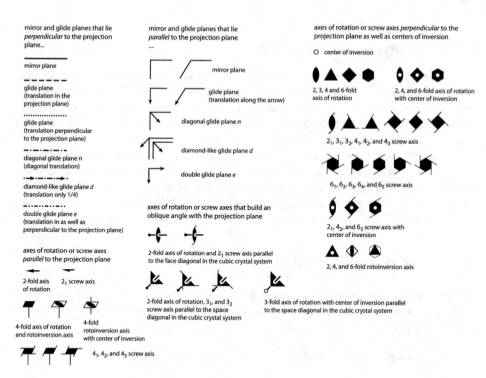

Fig. 6.26 The most important graphical symbols for symmetry elements used in the International Tables

the sake of simplicity, which do not quite conform with the use of the International Tables. For example, a two-fold axis of rotation running *in* the projection plane should not be marked by an axis at the end of which there is a correspondingly horizontally or vertically perspective-distorted ellipse; instead, it is better to use arrows. The type of dashing or dotting of a line also distinguishes whether a glide plane has a translational component that lies *in* or *perpendicular* to the projection plane. Furthermore, the really complicated symmetry element diagrams of space groups of the hexagonal or cubic crystal system use symbols that have not yet been discussed. This is partly due to the fact that in this case symmetry elements are not located *in* or *perpendicular* to projection planes but form an oblique angle with them. Additionally, a special graphical representation is used if several symmetry elements lie on top of each other. Therefore, a general overview of all graphical representations of symmetry elements used in the International Tables is given here, with the exception of the oblique mirrors and glide planes in the cubic space groups of crystal class $\bar{4}3m$ and $m\bar{3}m$ (**Fig. 6.26**).

6.4 About the Term Group: Brief Excursion into Group Theory

Now you already know almost everything about crystallographic point and space groups, but why the point group is called a point *group* and the space group space *group* is still unclear. The explanation for this is that the symmetry operations of a point or space group fulfill the mathematical properties of a group. The mathematical subsection dealing with the properties of groups is the group theory, which in turn represents a subfield of algebra. The group theory is highly complex and abstract. But understanding what a group *is*, is actually quite simple.

A set of elements, for example, pieces of fruit, integers, or symmetry operations, is a group if and only if that set satisfies the three group axioms, that is, certain conditions, with respect to a particular operator (peeling, summation, sequential application):

1. The associative law is satisfied.
2. There is a neutral element.
3. There is an inverse element to each element.

This can be illustrated using the example of integers (... −3, −2, −1, 0, +1, +2, +3...) and the operator addition.

1. The associative law refers to a set of (at least) three elements and says that the elements may be connected with parentheses as you like. For example, specifically for the operation addition, it applies:

$$3 + \left[(-7) + 5 \right] = \left[3 + (-7) \right] + 5 = 1$$

   However, with regard to the linkage of three elements, it also follows that we can omit the parentheses and the result is independent of the order of the linkage.
2. The neutral element of the integers is zero: it leaves all other members of the group unchanged $(+7 + 0 = +7)$. Thus, the second axiom is also fulfilled.
3. The fact that there is an inverse element for each element, i.e., for each integer, is also immediately obvious. 3 is the inverse element to −3, and −49 is the inverse element to +49. If an element is linked to the inverse element, i.e., added here, the neutral element results $(-368 + 368 = 0)$.

It also applies to a group that it is complete with respect to the operator. That means, that no new element may appear when the operator is executed; otherwise, the group was not complete. For the addition as an operator, by the way as well as for subtraction, and the group of integers, this also holds: any addition or subtraction always produces an element that is already in the group, an integer. Conversely, the integers do not form a group with respect to the operator division, because $5/2 = 2.5$ – not an integer!

Transferred to the point or space groups, here the symmetry operations (not the symmetry elements!) form the elements of the group. The operator is simply the execution or consecutive execution of these symmetry operations. Let's check, e.g., on the basis of the space group *Pmm2*, whether the group axioms are fulfilled:

1. The sequence of consecutive execution of three symmetry operations does not matter. For example, if I mirror a motif first at the mirror plane of the first viewing direction in the orthorhombic crystal system (perpendicular to *a*), then mirror it at the mirror plane of the second viewing direction (perpendicular to *b*), and finally perform a rotation about the two-fold axis of rotation parallel to the *c* direction, or if I do it the other way round, the motif will arrive at an identical point after these three operations.

   Remember, however, that the commutative law does not necessarily have to apply: the result of *two* successive symmetry operations may depend on their order.
2. There is a neutral element: that is identity (= doing nothing)!
3. And there is also an inverse element for each element (= symmetry operation). Even if it may seem like a trick to you, in this case, each element forms its own inverse element!

Well, the condition of the axiom was that when linking an element with its inverse than the result will be the neutral element. And this is the case: for example, if we have performed a mirroring on a plane perpendicular to *a* and then again on the same plane, the motif lands where it was at the starting point. The result would be identical with doing nothing, i.e., applying the identity operation. It is also the case for the rotation by 180°: applying the two-fold axis of rotation two times transfers a point to its initial position. And it also applies for the translation along the primitive lattice: the translation along the *a* lattice vector a converts the point with coordinates 0,0,0 into coordinates 1,0,0 – but this point is identical to the point at 0,0,0.

The fact that crystallographic point and space groups are groups in mathematical terms may also explain why many mathematicians have a penchant for crystallography. As beautiful as crystals are, they are also highly mathematical objects.

6.5 The SpaceGroupVisualizer

The International Tables – useful as they are – have one decisive disadvantage: they represent the space groups, the general positions, and the type and position of the symmetry elements only by means of two-dimensional projections. For those of the readers who want to gain a three-dimensional insight into the symmetry frameworks of all 230 space groups, there is a great software project called "SpaceGroupVisualizer" (■ Fig. 6.27). In this tool, which is also capable of generating a red/green or magenta/cyan stereo view, the unit cells with the symmetry frameworks can be freely rotated in 3D. But it offers even more: if you touch one of the symmetry elements with the mouse pointer, two general positions, which are represented by cubes with two capped cones perpendicular to each other, are dynamically transferred into each other according to the symmetry operation of the symmetry element. This software is an ideal supplement to the ITA. More information and the download link can be found at this URL: ▶ http://spacegroup.info/

■ **Fig. 6.27** Three-dimensional visualization of the symmetry framework with the "SpaceGroupVisualizer" software, here using the space group *P2/m* as an example

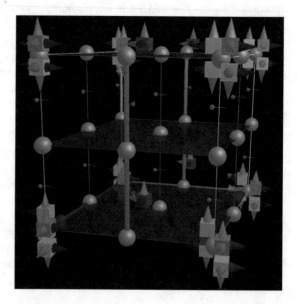

6.6 Tabular Overview of the 230 Space Groups

This chapter concludes with a complete tabular overview of all 230 space groups, in which the space group number, the long and short symbol, the crystal class, and to which crystal system the space groups belong are indicated (◘ Table 6.3).

◘ **Table 6.3** Overview of space groups with information on the space group number, the long and short symbol, the crystal class, and to which the crystal system the space groups belong

Crystal system	No.	Short symbol	Long symbol	Crystal class
Triclinic	1	$P1$	$P1$	1
	2	$P\bar{1}$	$P\bar{1}$	$\bar{1}$
Monoclinic	3	$P2$	$P2$	2
	4	$P2_1$	$P12_11$	2
	5	$C2$	$C121$	2
	6	Pm	$P1m1$	m
	7	Pc	$P1c1$	m
	8	Cm	$C1m1$	m
	9	Cc	$C1c1$	m
	10	$P\dfrac{2}{m}$	$P1\dfrac{2}{m}1$	$2/m$
	11	$P\dfrac{2_1}{m}$	$P1\dfrac{2_1}{m}1$	$2/m$
	12	$C\dfrac{2}{m}$	$C1\dfrac{2}{m}1$	$2/m$
	13	$P\dfrac{2}{c}$	$P1\dfrac{2}{c}1$	$2/m$
	14	$P\dfrac{2_1}{c}$	$P1\dfrac{2_1}{c}1$	$2/m$
	15	$C\dfrac{2}{c}$	$C1\dfrac{2}{c}1$	$2/m$
Orthorhombic	16	$P222$	$P222$	222
	17	$P222_1$	$P222_1$	222
	18	$P2_12_12$	$P2_12_12$	222
	19	$P2_12_12_1$	$P2_12_12_1$	222
	20	$C222_1$	$C222_1$	222
	21	$C222$	$C222$	222
	22	$F222$	$F222$	222
	23	$I222$	$I222$	222
	24	$I2_12_12_1$	$I2_12_12_1$	222

(continued)

☐ Table 6.3 (continued)

Crystal system	No.	Short symbol	Long symbol	Crystal class
	25	$Pmm2$	$Pmm2$	$mm2$
	26	$Pmc2_1$	$Pmc2_1$	$mm2$
	27	$Pcc2$	$Pcc2$	$mm2$
	28	$Pma2$	$Pma2$	$mm2$
	29	$Pca2_1$	$Pca2_1$	$mm2$
	30	$Pnc2$	$Pnc2$	$mm2$
	31	$Pmn2_1$	$Pmn2_1$	$mm2$
	32	$Pba2$	$Pba2$	$mm2$
	33	$Pna2_1$	$Pna2_1$	$mm2$
	34	$Pnn2$	$Pnn2$	$mm2$
	35	$Cmm2$	$Cmm2$	$mm2$
	36	$Cmc2_1$	$Cmc2_1$	$mm2$
	37	$Ccc2$	$Ccc2$	$mm2$
	38	$Amm2$	$Amm2$	$mm2$
	39	$Aem2$	$Aem2$	$mm2$
	40	$Ama2$	$Ama2$	$mm2$
	41	$Aea2$	$Aea2$	$mm2$
	42	$Fmm2$	$Fmm2$	$mm2$
	43	$Fdd2$	$Fdd2$	$mm2$
	44	$Imm2$	$Imm2$	$mm2$
	45	$Iba2$	$Iba2$	$mm2$
	46	$Ima2$	$Ima2$	$mm2$
	47	$Pmmm$	$P\dfrac{2}{m}\dfrac{2}{m}\dfrac{2}{m}$	mmm
	48	$Pnnn$	$P\dfrac{2}{n}\dfrac{2}{n}\dfrac{2}{n}$	mmm
	49	$Pccm$	$P\dfrac{2}{c}\dfrac{2}{c}\dfrac{2}{m}$	mmm
	50	$Pban$	$P\dfrac{2}{b}\dfrac{2}{a}\dfrac{2}{n}$	mmm

▣ **Table 6.3** (continued)

Crystal system	No.	Short symbol	Long symbol	Crystal class
	51	Pmma	$P\dfrac{2_1}{m}\dfrac{2}{m}\dfrac{2}{m}$	mmm
	52	Pnna	$P\dfrac{2}{n}\dfrac{2}{n}\dfrac{2_1}{n}$	mmm
	53	Pmna	$P\dfrac{2}{m}\dfrac{2}{n}\dfrac{2_1}{a}$	mmm
	54	Pcca	$P\dfrac{2_1}{c}\dfrac{2}{c}\dfrac{2}{a}$	mmm
	55	Pbam	$P\dfrac{2_1}{b}\dfrac{2_1}{a}\dfrac{2}{m}$	mmm
	56	Pccn	$P\dfrac{2_1}{c}\dfrac{2_1}{c}\dfrac{2}{n}$	mmm
	57	Pbcm	$P\dfrac{2_1}{b}\dfrac{2}{c}\dfrac{2_1}{m}$	mmm
	58	Pnnm	$P\dfrac{2_1}{n}\dfrac{2_1}{n}\dfrac{2}{m}$	mmm
	59	Pmmn	$P\dfrac{2_1}{m}\dfrac{2_1}{m}\dfrac{2}{n}$	mmm
	60	Pbcn	$P\dfrac{2_1}{b}\dfrac{2}{c}\dfrac{2_1}{n}$	mmm
	61	Pbca	$P\dfrac{2_1}{b}\dfrac{2_1}{c}\dfrac{2_1}{a}$	mmm
	62	Pnma	$P\dfrac{2_1}{n}\dfrac{2_1}{m}\dfrac{2_1}{a}$	mmm
	63	Cmcm	$C\dfrac{2}{m}\dfrac{2}{c}\dfrac{2_1}{m}$	mmm
	64	Cmca	$C\dfrac{2}{m}\dfrac{2}{c}\dfrac{2_1}{a}$	mmm
	65	Cmmm	$C\dfrac{2}{m}\dfrac{2}{m}\dfrac{2}{m}$	mmm

(continued)

■ Table 6.3 (continued)

Crystal system	No.	Short symbol	Long symbol	Crystal class
	66	Cccm	$C\dfrac{2\,2\,2}{c\,c\,m}$	mmm
	67	Cmme	$C\dfrac{2\,2\,2}{m\,m\,e}$	mmm
	68	Ccce	$C\dfrac{2\,2\,2}{c\,c\,e}$	mmm
	69	Fmmm	$F\dfrac{2\,2\,2}{m\,m\,m}$	mmm
	70	Fddd	$F\dfrac{2\,2\,2}{d\,d\,d}$	mmm
	71	Immm	$I\dfrac{2\,2\,2}{m\,m\,m}$	mmm
	72	Ibam	$I\dfrac{2\,2\,2}{b\,a\,m}$	mmm
	73	Ibca	$I\dfrac{2\,2\,2}{b\,c\,a}$	mmm
	74	Imma	$I\dfrac{2\,2\,2}{m\,m\,a}$	mmm
Tetragonal	75	P4	P4	4
	76	$P4_1$	$P4_1$	4
	77	$P4_2$	$P4_2$	4
	78	$P4_3$	$P4_3$	4
	79	I4	I4	4
	80	$I4_1$	$I4_1$	4
	81	$P\bar{4}$	$P\bar{4}$	$\bar{4}$
	82	$I\bar{4}$	$I\bar{4}$	$\bar{4}$
	83	$P\dfrac{4}{m}$	$P\dfrac{4}{m}$	$\dfrac{4}{m}$
	84	$P\dfrac{4_2}{m}$	$P\dfrac{4_2}{m}$	$\dfrac{4}{m}$

Table 6.3 (continued)

Crystal system	No.	Short symbol	Long symbol	Crystal class
	85	$P\dfrac{4}{n}$	$P\dfrac{4}{n}$	$\dfrac{4}{m}$
	86	$P\dfrac{4_2}{n}$	$P\dfrac{4_2}{n}$	$\dfrac{4}{m}$
	87	$I\dfrac{4}{m}$	$I\dfrac{4}{m}$	$\dfrac{4}{m}$
	88	$I\dfrac{4_1}{a}$	$I\dfrac{4_1}{a}$	$\dfrac{4}{m}$
	89	$P422$	$P422$	422
	90	$P42_12$	$P42_12$	422
	91	$P4_122$	$P4_122$	422
	92	$P4_12_12$	$P4_12_12$	422
	93	$P4_222$	$P4_222$	422
	94	$P4_22_12$	$P4_22_12$	422
	95	$P4_322$	$P4_322$	422
	96	$P4_32_12$	$P4_32_12$	422
	97	$I422$	$I422$	422
	98	$I4_122$	$I4_122$	422
	99	$P4mm$	$P4mm$	$4mm$
	100	$P4bm$	$P4bm$	$4mm$
	101	$P4_2cm$	$P4_2cm$	$4mm$
	102	$P4_2nm$	$P4_2nm$	$4mm$
	103	$P4cc$	$P4cc$	$4mm$
	104	$P4nc$	$P4nc$	$4mm$
	105	$P4_2mc$	$P4_2mc$	$4mm$
	106	$P4_2bc$	$P4_2bc$	$4mm$
	107	$I4mm$	$I4mm$	$4mm$
	108	$I4cm$	$I4cm$	$4mm$
	109	$I4_1md$	$I4_1md$	$4mm$
	110	$I4_1cd$	$I4_1cd$	$4mm$

(continued)

□ Table 6.3 (continued)

Crystal system	No.	Short symbol	Long symbol	Crystal class
	111	$P\bar{4}2m$	$P\bar{4}2m$	$\bar{4}2m$
	112	$P\bar{4}2c$	$P\bar{4}2c$	$\bar{4}2m$
	113	$P\bar{4}2_1m$	$P\bar{4}2_1m$	$\bar{4}2m$
	114	$P\bar{4}2_1c$	$P\bar{4}2_1c$	$\bar{4}2m$
	115	$P\bar{4}m2$	$P\bar{4}m2$	$\bar{4}2m$
	116	$P\bar{4}c2$	$P\bar{4}c2$	$\bar{4}2m$
	117	$P\bar{4}b2$	$P\bar{4}b2$	$\bar{4}2m$
	118	$P\bar{4}n2$	$P\bar{4}n2$	$\bar{4}2m$
	119	$I\bar{4}m2$	$I\bar{4}m2$	$\bar{4}2m$
	120	$I\bar{4}c2$	$I\bar{4}c2$	$\bar{4}2m$
	121	$I\bar{4}2m$	$I\bar{4}2m$	$\bar{4}2m$
	122	$I\bar{4}2d$	$I\bar{4}2d$	$\bar{4}2m$
	123	$P\frac{4}{m}mm$	$P\frac{4}{m}\frac{2}{m}\frac{2}{m}$	$\frac{4}{m}mm$
	124	$P\frac{4}{m}cc$	$P\frac{4}{m}\frac{2}{c}\frac{2}{c}$	$\frac{4}{m}mm$
	125	$P\frac{4}{n}bm$	$P\frac{4}{n}\frac{2}{b}\frac{2}{m}$	$\frac{4}{m}mm$
	126	$P\frac{4}{n}nc$	$P\frac{4}{n}\frac{2}{n}\frac{2}{c}$	$\frac{4}{m}mm$
	127	$P\frac{4}{m}bm$	$P\frac{4}{m}\frac{2_1}{b}\frac{2}{m}$	$\frac{4}{m}mm$
	128	$P\frac{4}{m}nc$	$P\frac{4}{m}\frac{2_1}{n}\frac{2}{c}$	$\frac{4}{m}mm$

◻ **Table 6.3** (continued)

Crystal system	No.	Short symbol	Long symbol	Crystal class
	129	$P\dfrac{4}{n}mm$	$P\dfrac{4}{n}\dfrac{2_1}{m}\dfrac{2}{m}$	$\dfrac{4}{m}mm$
	130	$P\dfrac{4}{n}cc$	$P\dfrac{4}{n}\dfrac{2_1}{c}\dfrac{2}{c}$	$\dfrac{4}{m}mm$
	131	$P\dfrac{4_2}{m}mc$	$P\dfrac{4_2}{m}\dfrac{2}{m}\dfrac{2}{c}$	$\dfrac{4}{m}mm$
	132	$P\dfrac{4_2}{m}cm$	$P\dfrac{4_2}{m}\dfrac{2}{c}\dfrac{2}{m}$	$\dfrac{4}{m}mm$
	133	$P\dfrac{4_2}{n}bc$	$P\dfrac{4_2}{n}\dfrac{2}{b}\dfrac{2}{c}$	$\dfrac{4}{m}mm$
	134	$P\dfrac{4_2}{n}nm$	$P\dfrac{4_2}{n}\dfrac{2}{n}\dfrac{2}{m}$	$\dfrac{4}{m}mm$
	135	$P\dfrac{4_2}{m}bc$	$P\dfrac{4_2}{m}\dfrac{2_1}{b}\dfrac{2}{c}$	$\dfrac{4}{m}mm$
	136	$P\dfrac{4_2}{m}nm$	$P\dfrac{4_2}{m}\dfrac{2_1}{n}\dfrac{2}{m}$	$\dfrac{4}{m}mm$
	137	$P\dfrac{4_2}{n}mc$	$P\dfrac{4_2}{n}\dfrac{2_1}{m}\dfrac{2}{c}$	$\dfrac{4}{m}mm$
	138	$P\dfrac{4_2}{n}cm$	$P\dfrac{4_2}{n}\dfrac{2_1}{c}\dfrac{2}{m}$	$\dfrac{4}{m}mm$
	139	$I\dfrac{4}{m}mm$	$I\dfrac{4}{m}\dfrac{2}{m}\dfrac{2}{m}$	$\dfrac{4}{m}mm$
	140	$I\dfrac{4}{m}cm$	$I\dfrac{4}{m}\dfrac{2}{c}\dfrac{2}{m}$	$\dfrac{4}{m}mm$
	141	$I\dfrac{4_1}{a}md$	$I\dfrac{4_1}{a}\dfrac{2}{m}\dfrac{2}{d}$	$\dfrac{4}{m}mm$
	142	$I\dfrac{4_1}{a}cd$	$I\dfrac{4_1}{a}\dfrac{2}{c}\dfrac{2}{d}$	$\dfrac{4}{m}mm$

(continued)

6

◘ **Table 6.3** (continued)

Crystal system	No.	Short symbol	Long symbol	Crystal class
Trigonal	143	$P3$	$P3$	3
	144	$P3_1$	$P3_1$	3
	145	$P3_2$	$P3_2$	3
	146	$R3$	$R3$	3
	147	$P\bar{3}$	$P\bar{3}$	$\bar{3}$
	148	$R\bar{3}$	$R\bar{3}$	$\bar{3}$
	149	$P312$	$P312$	32
	150	$P321$	$P321$	32
	151	$P3_112$	$P3_112$	32
	152	$P3_121$	$P3_121$	32
	153	$P3_212$	$P3_212$	32
	154	$P3_221$	$P3_221$	32
	155	$R32$	$R32$	32
	156	$P3m1$	$P3m1$	$3m$
	157	$P31m$	$P31m$	$3m$
	158	$P3c1$	$P3c1$	$3m$
	159	$P31c$	$P31c$	$3m$
	160	$R3m$	$R3m$	$3m$
	161	$R3c$	$R3c$	$3m$
	162	$P\bar{3}1m$	$P\bar{3}1\frac{2}{m}$	$\bar{3}m$
	163	$P\bar{3}1c$	$P\bar{3}1\frac{2}{c}$	$\bar{3}m$
	164	$P\bar{3}m1$	$P\bar{3}\frac{2}{m}1$	$\bar{3}m$
	165	$P\bar{3}c1$	$P\bar{3}\frac{2}{c}1$	$\bar{3}m$
	166	$R\bar{3}m$	$R\bar{3}\frac{2}{m}$	$\bar{3}m$
	167	$R\bar{3}c$	$R\bar{3}\frac{2}{c}$	$\bar{3}m$

◻ **Table 6.3** (continued)

Crystal system	No.	Short symbol	Long symbol	Crystal class
Hexagonal	168	$P6$	$P6$	6
	169	$P6_1$	$P6_1$	6
	170	$P6_5$	$P6_5$	6
	171	$P6_2$	$P6_2$	6
	172	$P6_4$	$P6_4$	6
	173	$P6_3$	$P6_3$	6
	174	$P\bar{6}$	$P\bar{6}$	$\bar{6}$
	175	$P\dfrac{6}{m}a$	$P\dfrac{6}{m}$	$\dfrac{6}{m}$
	176	$P\dfrac{6_3}{m}$	$P\dfrac{6_3}{m}$	$\dfrac{6}{m}$
	177	$P622$	$P622$	622
	178	$P6_122$	$P6_122$	622
	179	$P6_522$	$P6_522$	622
	180	$P6_222$	$P6_222$	622
	181	$P6_422$	$P6_422$	622
	182	$P6_322$	$P6_322$	622
	183	$P6mm$	$P6mm$	$6mm$
	184	$P6cc$	$P6cc$	$6mm$
	185	$P6_3cm$	$P6_3cm$	$6mm$
	186	$P6_3mc$	$P6_3mc$	$6mm$
	187	$P\bar{6}m2$	$P\bar{6}m2$	$\bar{6}m2a$
	188	$P\bar{6}c2$	$P\bar{6}c2$	$\bar{6}m2$
	189	$P\bar{6}2m$	$P\bar{6}2m$	$\bar{6}m2$
	190	$P\bar{6}2c$	$P\bar{6}2c$	$\bar{6}m2$
	191	$P\dfrac{6}{m}mm$	$P\dfrac{6\;2\;2}{m\,m\,m}$	$\dfrac{6}{m}mm$
	192	$P\dfrac{6}{m}cc$	$P\dfrac{6\;2\;2}{m\,c\,c}$	$\dfrac{6}{m}mm$

(continued)

Table 6.3 (continued)

Crystal system	No.	Short symbol	Long symbol	Crystal class
	193	$P\dfrac{6_3}{m}cm$	$P\dfrac{6_3}{m}\dfrac{2}{c}\dfrac{2}{m}$	$\dfrac{6}{m}mm$
	194	$P\dfrac{6_3}{m}mc$	$P\dfrac{6_3}{m}\dfrac{2}{m}\dfrac{2}{c}$	$\dfrac{6}{m}mm$
Cubic	195	$P23$	$P23$	23
	196	$F23$	$F23$	23
	197	$I23$	$I23$	23
	198	$P2_13$	$P2_13$	23
	199	$I2_13$	$I2_13$	23
	200	$Pm\bar{3}$	$P\dfrac{2}{m}\bar{3}$	$m\bar{3}$
	201	$Pn\bar{3}$	$P\dfrac{2}{n}\bar{3}$	$m\bar{3}$
	202	$Fm\bar{3}$	$F\dfrac{2}{m}\bar{3}$	$m\bar{3}$
	203	$Fd\bar{3}$	$F\dfrac{2}{d}\bar{3}$	$m\bar{3}$
	204	$Im\bar{3}$	$I\dfrac{2}{m}\bar{3}$	$m\bar{3}$
	205	$Pa\bar{3}$	$P\dfrac{2_1}{a}\bar{3}$	$m\bar{3}$
	206	$Ia\bar{3}$	$I\dfrac{2_1}{a}\bar{3}$	$m\bar{3}$
	207	$P432$	$P432$	432
	208	$P4_232$	$P4_232$	432
	209	$F432$	$F432$	432
	210	$F4_132$	$F4_132$	432
	211	$I432$	$I432$	432
	212	$P4_332$	$P4_332$	432
	213	$P4_132$	$P4_132$	432
	214	$I4_132$	$I4_132$	432

Table 6.3 (continued)

Crystal system	No.	Short symbol	Long symbol	Crystal class
	215	$P\bar{4}3m$	$P\bar{4}3m$	$\bar{4}3m$
	216	$F\bar{4}3m$	$F\bar{4}3m$	$\bar{4}3m$
	217	$I\bar{4}3m$	$I\bar{4}3m$	$\bar{4}3m$
	218	$P\bar{4}3n$	$P\bar{4}3n$	$\bar{4}3m$
	219	$F\bar{4}3c$	$F\bar{4}3c$	$\bar{4}3m$
	220	$I\bar{4}3d$	$I\bar{4}3d$	$\bar{4}3m$
	221	$Pm\bar{3}m$	$P\dfrac{4}{m}\bar{3}\dfrac{2}{m}$	$m\bar{3}m$
	222	$Pn\bar{3}n$	$P\dfrac{4}{n}\bar{3}\dfrac{2}{n}$	$m\bar{3}m$
	223	$Pm\bar{3}n$	$P\dfrac{4_2}{m}\bar{3}\dfrac{2}{n}$	$m\bar{3}m$
	224	$Pn\bar{3}m$	$P\dfrac{4_2}{n}\bar{3}\dfrac{2}{m}$	$m\bar{3}m$
	225	$Fm\bar{3}m$	$F\dfrac{4}{m}\bar{3}\dfrac{2}{m}$	$m\bar{3}m$
	226	$Fm\bar{3}c$	$F\dfrac{4}{m}\bar{3}\dfrac{2}{c}$	$m\bar{3}m$
	227	$Fd\bar{3}m$	$F\dfrac{4_1}{d}\bar{3}\dfrac{2}{m}$	$m\bar{3}m$
	228	$Fd\bar{3}c$	$F\dfrac{4_1}{d}\bar{3}\dfrac{2}{c}$	$m\bar{3}m$
	229	$Im\bar{3}m$	$I\dfrac{4}{m}\bar{3}\dfrac{2}{m}$	$m\bar{3}m$
	230	$Ia\bar{3}d$	$I\dfrac{4_1}{a}\bar{3}\dfrac{2}{d}$	$m\bar{3}m$

References

1. Hahn T (ed) (2016) International tables for crystallography, Vol. A, Space-group symmetry. 6th edn. Wiley, New York
2. A hypertext book of crystallographic space group diagrams and tables. © Copyright 1997–1999. Birkbeck College, University of London. http://img.chem.ucl.ac.uk/sgp/mainmenu.htm. Accessed 30 Sept 2019
3. Wyckoff RWG (1922) The analytical expression of the results of the theory of space groups. Publication no. 318. Carnegie Institution of Washington, Washington
4. Urusov VS, Nadezhina TN (2009) Frequency distribution and selection of space groups in inorganic crystal chemistry. J Struct Chem 50(Suppl):S22–S37. https://doi.org/10.1007/s10947-009-0186-9

6

Some Real Crystal Structures: From Theory to Practice

© Springer Nature Switzerland AG 2020
F. Hoffmann, *Introduction to Crystallography*, https://doi.org/10.1007/978-3-030-35110-6_7

In the previous chapter, we had to deal with plenty of theory. All the time we talked about space groups without mentioning even a single crystal structure. You could almost have been left with the impression that this is a book about mathematical or geometric crystallography. This should not be the case – so it's high time to return to real crystal structures!

In this chapter, we will discuss some selected crystal structures to deepen our newly acquired theoretical knowledge. The choice of examples does not necessarily reflect the "importance" of crystals or minerals, though some are very common, naturally occurring species, such as saline or graphite. Rather, it is about linking them to theoretical considerations and making some concepts more tangible on the basis of real structures.

7.1 Ethylene and the Visualization of Symmetry Elements with the Software Mercury

Ethene (also called ethylene) is a molecule belonging to the class of so-called unsaturated hydrocarbons and has the formula C_2H_4. It has a double bond between the two carbon atoms. It is one of the most commonly produced basic organic chemicals and is used, for instance, for the production of the important thermoplastic polyethylene.

The molecular shape of ethene is planar, i.e., all atoms are in one plane. The bond angles are approximately 120°. The structure and a ball-and-stick model of ethene are shown in ◘ Fig. 7.1. At room temperature, ethene is a colorless gas, but when cooled down to −103.7 °C, it condenses into a liquid. At temperatures below −169.2 °C, ethene finally crystallizes in the space group $P2_1/n$ with the lattice parameters: $a = 4.626$ Å, $b = 6.620$ Å, $c = 4.067$ Å, $\alpha = 90°$, $\beta = 94.39°$, and $\gamma = 90°$.

Let's take a look at the crystal structure. We could do that with the VESTA program, but in this case, we want to use the Mercury program, which is provided as freeware by the Cambridge Crystallographic Data Center (CCDC) and is available for all major operating systems. You can find the link to download at [1] or search the Internet with the keywords "CCDC" and "Mercury." This software has an invaluable advantage compared to other programs for the visualization of crystal structures: in addition to the representation of the crystal structure, Mercury is also able to display the symmetry elements of the space group or to superimpose it on the structure. Thus, the program is a great tool to develop a deeper understanding of the symmetry relationships of the parts of a concrete crystal structure and to visualize the positions of the symmetry elements in the unit cell.

The CIF file of ethene is available on the accompanying page of the book at ▶ https://crystalsymmetry.wordpress.com/textbook/. After downloading and installing Mercury, open the CIF file. You should see an ethene molecule as a wireframe against a black background; this is the basic view of Mercury. In the header area of the control panels, there is a drop-down menu under "Manage Styles ..." to the right, where a more representative style can be selected (◘ Fig. 7.2a); alternatively, the style can be defined via the menu item "Display – Styles" (◘ Fig. 7.2b) and the background color via "Display – Colors – Background settings..." (◘ Fig. 7.2c).

◘ **Fig. 7.1** The structure of the molecule ethene with molecular formula C_2H_4 (left) and a ball-and-stick representation (right); C = gray, H = white

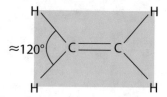

Fig. 7.2 (a–c) Menu and dialog boxes for setting different display styles in the software Mercury

To visualize the molecular packing of the crystal structure and the unit cell, please check-mark the option "Packing" at the panel "Display Options" in the lower left corner (■ Fig. 7.3).

Mouse movements while holding down the left mouse button allow to rotate the structure three-dimensionally, and clicking and dragging with the right mouse button control the zoom level. You should now see a similar picture in front of you as shown in ■ Fig. 7.4; if your cell axes are not colored, a right-click on one of the axes and choosing "Color Unit Cell Axes" in the pop-up selection menu will help.

There is an ethene molecule at each corner and in the center of the unit cell, but the ethene molecule in the center has a different orientation, so it is not a body-centered, but a primitive lattice, consistent with the space group symbol.

7.1.1 Center of Inversion and the Asymmetric Unit

Now we can superimpose the existing symmetry elements on the structure by selecting the menu item "Display – Symmetry Elements …." In the dialog box (■ Fig. 7.5),

Fig. 7.3   Control panel in Mercury to visualize the packing of the crystal structure and not just a single molecule

Fig. 7.4   The unit cell of ethylene viewed in Mercury. Note that the molecule in the center of the cell has a different orientation compared to the molecules at the corner positions

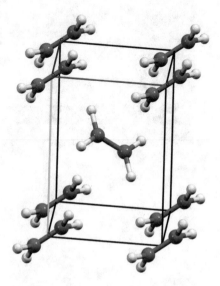

7

Fig. 7.5   Dialog box for visualizing the symmetry elements of a crystal structure in the software Mercury. Leave only the check mark at the inversion centers, and deselect all other symmetry elements. Adjust the "Colour" of the center of inversions and their "Size" with the slider

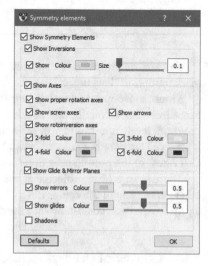

◘ Fig. 7.6 Crystal structure of ethylene with additional superposition of the inversion centers in the unit cell

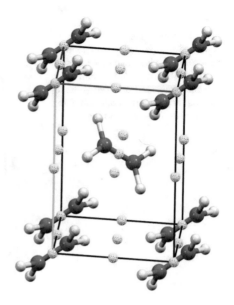

all symmetry elements are initially selected by default, but it is clearer to consider each kind separately. Let's start with the inversion centers.

It can be seen that inversion centers are located at all corners, at all edge bisectors, and at all face centers. Additionally, one is located in the center of the cell (◘ Fig. 7.6).

The origin of the cell lies at an inversion center, which is in accordance with the rules. This inversion center is transformed by the translational vectors of the lattice to all other vertices of the unit cell. Incidentally, this also automatically results in the inversion centers at the edge bisectors and face centers.

Interestingly, in the case of ethene, the molecule itself also has an inversion center. If we now look at the ethene molecules at their locations in the crystal structure, we realize that the inversion center of the ethene coincides with the position of the inversion centers of the cell! This applies both to the ethene molecules at the corners and to the molecule in the center. This means that the crystallographic inversion center generates the other half from one half of the ethene molecule when this point mirroring is carried out (◘ Fig. 7.7).

What does this mean for the asymmetric unit? We have learned that the asymmetric unit must not have eigensymmetry (see ▶ Chap. 6). Exactly this would be the case if the whole molecule formed the asymmetric unit. Thus, because the inversion center of the molecule coincides with the inversion center of the space group, the inversion center of the molecule is said to be crystallographically effective in such a way that only one half of the molecule forms the asymmetric unit. As a reminder, the asymmetric unit is the part of the structure that gives the complete crystal structure when applying all of the symmetry elements of the space group. There are two ways to verify that the asymmetric unit actually consists of only half of the molecule:

- The first option is to do this within Mercury. You can do so by first unchecking the "Packing" option in the lower left corner of the window under "Display Options" and

Fig. 7.7 The centers of inversion are not only those of the unit cell: they coincide with those of the ethylene molecules. This is shown here for the central ethene molecule. A point reflection transfers both hydrogen atoms and the carbon atom of one half of the molecule into the atoms of the other half

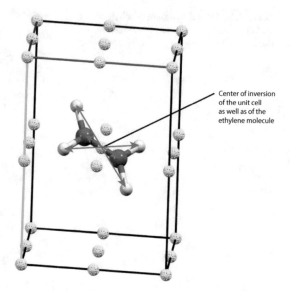

Center of inversion of the unit cell as well as of the ethylene molecule

Fig. 7.8 In the control panel "Display Options" at the bottom of the screen, it is possible to show the asymmetric unit. First deselect "Packing" and then check the box for "Asymmetric Unit"

then setting it to "Asymmetric Unit" (■ Fig. 7.8). In fact, you should only be able to see half of the ethene molecule on your screen.

— The second possibility is to open the CIF file with a text editor and look at the section in which the fractional coordinates are stored – because in the CIF file, only the atoms of the asymmetric unit are specified, while the representation of one or more unit cells is done by the internal application of the symmetry elements. The corresponding three lines look like this:

— C1 C −0.11656   0.05382 −0.04075
— H1 H −0.18690   0.16980   0.11850
— H2 H −0.24300   0.02870 −0.26890

Thus, two hydrogen atoms and one carbon atom suffice as an asymmetric unit. To complete the relationship between the eigensymmetry of the motif and the symmetry of the space group, another aspect can be important, because you may object: isn't it the case that the ethylene molecule has more symmetry elements in addition to the inversion center? And also the three atoms of the asymmetric unit do not look completely asymmetric, as there is a mirror plane along the C-C bond axis (see ■ Fig. 7.1). That is correct – the asymmetric unit of ethylene has the point group $mm2$ or, for those who are more familiar with Schoenflies nomenclature, C_{2h}. But, these symmetry elements are not crystallographically effective because they do not coincide with any symmetry element of the space group. The remaining symmetry elements of the

229   **7**

7.1 · Ethylene and the Visualization of Symmetry Elements with the Software ...

Fig. 7.9 The point symmetry elements of the asymmetric unit of ethylene, which as such has symmetry *mm*2 and thus belongs to the same point group as the water molecule, are not crystallographically effective

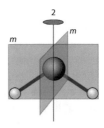

asymmetric unit, i.e., the two mirror planes and the two-fold axis of rotation (■ Fig. 7.9), find no correspondence to symmetry elements of the space group – neither of the type (in the space group $P2_1/n$ no mirror plane is present) nor of the orientation (a mirror plane of the molecule should have the same position and orientation as that of the space group). Thus, we can say that an asymmetric unit does not have to look completely asymmetrical.

Z and *Z'* Values

In crystallography, and in particular in X-ray diffraction analysis, it is common to specify the number of "formula units" per unit cell (= *Z*). In the case of molecular crystals, one formula unit corresponds to one molecule or the chemical formula of the molecule; in our example, this would be C_2H_4. So, we just ask how many molecules are contained in a unit cell. This number is always obtained by multiplying the multiplicity of the general position by the number of formula units which form the asymmetric unit (= *Z'*). In the case of ethylene, however, the asymmetric unit is only half a molecule (*Z'* = 0.5) because it is located at a special position. Thus, with the multiplicity of four, a *Z* value of two results. It should be noted that *Z* must be at least one, because less than one (whole) molecule per unit cell would be absurd, since the question would be where the rest of the molecule was. Conversely, there are also cases in which more than one molecule can form the asymmetric unit (*Z'* > 1).

There are some statistical studies concerning the frequency distributions of *Z* and *Z'* values for organic and organometallic compounds. The distribution, based on the approx. 400,000 crystal structures recorded up to 2006, results in values for *Z'* = 1 of 73%, for *Z'* < 1 of 16%, and for *Z'* > 1 of 11% [2]. Usually, this means in three out of four cases, exactly one molecule forms the asymmetric unit. Very much rarer, the molecule lies at a special position, and only half a molecule or a quarter of a molecule form the asymmetric unit. And even rarer, it happens that two or even more than two independent molecules together form the asymmetric unit (*Z'* ≥ 2). Values of more than two are almost rarities; their occurrence is clearly below 1% [3].

7.1.2 Further Symmetry Elements

Displaying the remaining symmetry elements will clarify further interrelationships of the crystal structure of ethene. In the "Display – Symmetry Elements ..." menu item, deselect the inversion centers, and set a check mark for the glide planes. You can also select a color of the plane and control the transparency level with a slider. As you can see, there are not only one, but two glide planes (perpendicular to the *b* direction), which run through the cell, one at the height $y = ¼$ and another at $y = ¾$. In the symmetry element diagram in the International Tables, only a single glide plane was drawn, labelled with ¼. In fact, the explanatory text accompanying the International Tables states that ¼ automatically refers to a further glide plane at $y = ¾$. In other cases, ⅛ implies an additional glide plane at ⅝ . You may already be able to derive why another glide plane at $y = ¾$ is present: the structure is centrosymmetric, i.e., in the center there is a center of inversion. This does not only affect the atoms, but also the other symmetry elements!

Or to put it another way, the center of inversion in the center of the cell can only exist if everything around it is arranged point symmetrically.

Because a cell is often quite stuffed, this usually makes it difficult to recognize the symmetry relationships. If you click on an ethylene molecule while holding down the *Shift* key, the entire molecule is marked, indicated by a yellow marking of the atoms (■ Fig. 7.10). Now you can hide these molecules with a right mouse click by selecting "Show/Hide – Hide atoms" in the pop-up menu (■ Fig. 7.11). Hide all molecules except one at the corner and the one in the center (according to ■ Fig. 7.12). Now you should be able to see more clearly how the glide plane connects these two molecules: the reflection of the central ethylene molecule at the glide plane inverts its orientation and moves the center of the molecule to the center of the face. Finally, since this is a diagonal glide plane *n*, the molecule has to be translated by ½ along the face diagonal so that it is mapped exactly on the molecule at the corner.

Since there is still a lot of imagination needed to fully comprehend geometric 3D operations, you will find a small animation for both the glide plane operation and the 2_1 screw axis movement on the companion page to the book on the Internet at ▶ https:// crystalsymmetry.wordpress.com/textbook.

7.2 Benzene and the Phenomenon of Polymorphism

On the one hand, this short section deals with the crystal structure of benzene; on the other hand, benzene serves as an example for the phenomenon of polymorphism, which will be briefly discussed here. Benzene has a very simple chemical shape, six carbon atoms

■ Fig. 7.10 By holding down the *Shift* key and clicking on an atom, the entire molecule is selected. Afterward the atoms can be hidden

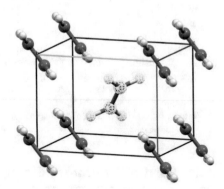

■ Fig. 7.11 A right-click after selecting atoms opens a pop-up menu that allows you to hide the atoms

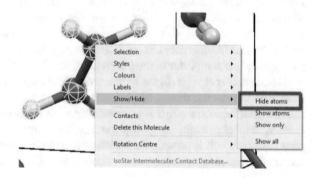

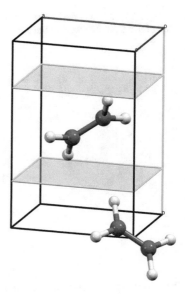

◘ Fig. 7.12　If the molecule in the center is first mirrored at the lower glide plane, its orientation is in accordance with the orientation of the molecule at the lower right corner of the cell. A subsequent diagonal shift transfers the two molecules into each other

◘ Fig. 7.13　The structure of benzene (C_6H_6) and a representation as ball-and-stick model

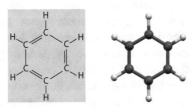

are connected to each other in the form of a regular hexagon, and each carbon atom carries another hydrogen atom (see ◘ Fig. 7.13); the chemical formula is C_6H_6.

Benzene belongs to the class of aromatic hydrocarbons and is – like ethylene – an extremely important industrial chemical from which numerous secondary products are synthesized. Unfortunately, it is also highly toxic and clearly carcinogenic, so when handling benzene, special precautions must be taken with regard to possible environmental and health hazards.

Benzene is a colorless liquid at room temperature, but below 5.5 °C, it solidifies and crystallizes in the space group *Pbca*. ◘ Figure 7.14 shows a so-called space-filling representation of the structure. Such representation has the disadvantage that you hardly see the wood for the trees, i.e., the crystal structure for the molecules. But it also has two advantages: first, it is a more realistic representation, because nature tends to avoid voids; it strives to pack matter as space-filling as possible. This is precisely the second advantage: even if nature fundamentally avoids making structures with holes, they do exist. And in a model that shows the atoms with their space demand that they have in nature, such empty voids will actually become identifiable.

But this is only one of several possible crystal structures of benzene. It only forms under certain conditions. So far, we have tacitly assumed that there is a kind of one-to-one

Fig. 7.14   A space-filling representation of the crystal packing of benzene

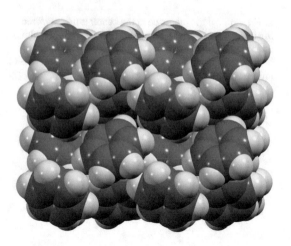

relationship: one substance, one structure. However, this only applies to certain pressure and temperature conditions, which should be specified for any given crystal structure. If no specifications are given, it implies either the crystal structure found at ambient conditions (i.e., at a pressure of approx. 1 bar and at room temperature) or the crystal structure which is formed when cooling down below the freezing point at normal pressure, which may be far below room temperature (as with ethylene). However, the specific crystal structure that is formed under certain conditions does not have to be the only possible crystal structure of a substance. Further cooling or change in pressure may cause the substance to transform from one into another crystal structure. This would be an example of a so-called solid-solid phase transition. In general, *physical* transformations of a substance – these can be changes in the state of aggregation or conversions from one structure to another in the solid state – are called phase transitions or phase transformations.

The phenomenon of polymorphism (gr. *poly* = many, gr. *morphé* = shape, form) describes the fact that one and the same substance can form different crystal structures. The individual different crystal structures of one and the same substance are called polymorphs (singular = the polymorph). This term may seem somewhat unfortunate, since polymorphs occasionally do in fact have different outer crystal shapes (see ▶ Sect. 2.1, Tracht and habit), but this is not a necessary condition.

Before we take a closer look at the phase diagram, let us consider two aspects of the structure of benzene in the space group *Pbca*, which belongs to the orthorhombic crystal system. Let's do it with the Mercury program. It's best to open the CIF file, provided on the Internet companion book page. ▪ Figure 7.15 shows a (slightly extended) unit cell in which the hydrogen atoms are omitted for the sake of clarity. Here, too, it may at first appear as if a centered cell was present, namely, an all-sided face-centered structure; after all, benzene molecules are present not only at all corners but also on all centers of the faces; but since they have a different orientation, this is actually just a primitive cell.

The second aspect concerns a variant of one of the symmetry theorems: two mirror planes perpendicular to each other automatically generate a two-fold axis of rotation along their cutting edge. Now the crystal structure of benzene does not have simple mirror planes but three glide planes, two of which are perpendicular to each other. Which other symmetry elements could be automatically present here? Quite clearly, the corresponding

◘ **Fig. 7.15** Unit cell of the crystal structure of benzene in the space group *Pbca* with the lattice parameters: $a = 7.390$ Å, $b = 9.420$ Å, $c = 6.810$ Å, $\alpha = \beta = \gamma = 90°$

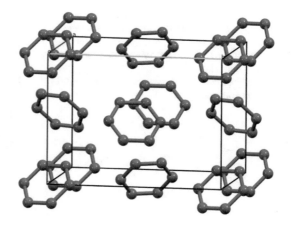

rotational symmetry elements with a translation component! And since the glide planes have a translation component of ½, it is clear that these must be 2_1 screw axes. Remember, however, that not only two glide planes must be perpendicular to each other to generate a 2_1 screw axis, but also the corresponding translation components. In the space group *Pbcn*, each two of the three glide planes are perpendicular to each other, but the direction of translation of the glide plane *n* runs diagonally, so here only two 2_1 screw axes are present.

You can check the presence of the screw axes by opening the CIF file again with the program Mercury and display the screw axes with the command "Display – Symmetry Elements ... Show screw axes." You can see that screw axes run parallel to all three crystallographic axes (see ◘ Fig. 7.16).

If you also display the glide planes, you can see that the screw axes run exactly along the cutting edges of two glide planes that are perpendicular to each other.

7.2.1 The Phase Diagram of Benzene

The crystal structure with the space group *Pbca* is not the only one possible for benzene. At constant temperature and increasing pressure, a phase transition into the polymorphs II (space group $P4_32_12$), III (space group $P2_1/c$), III' (also space group $P2_1/c$, but with different lattice parameters compared to III), and finally IV (space group *Pbam*) takes place. At higher temperatures under increased pressure, a polymorph V is additionally postulated, at very low temperatures and comparatively low pressures a polymorph I'. However, the existence and nature of the latter two phases have not yet been fully established experimentally. A preliminary schematic phase diagram of benzene is shown in ◘ Fig. 7.17 [4].

7.2.2 Crystal Structure Prediction: As Good or as Bad as the Weather Forecast?

The fact that a crystalline substance changes into another crystal structure at higher pressures may not come as a surprise. However, it is remarkable that even today it is still impossible for the vast majority of substances to predict which concrete crystal structure they will form under which conditions. And this although all basic physical or physico-

7

■ **Fig. 7.16**   In the short notation of the space group *Pbca*, the 2_1 screw axes (as well as the centers of inversions) are not explicitly mentioned; they are shown here in blue and are running along the intersecting edges of each two perpendicular glide planes. The long symbol takes these screw axes into account and reads $P\dfrac{2_1}{b}\dfrac{2_1}{c}\dfrac{2_1}{a}$

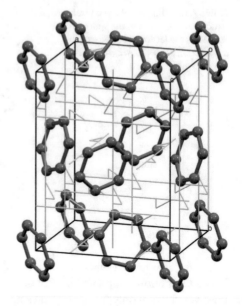

■ **Fig. 7.17**   The phase diagram of benzene, i.e., the structures or phases formed by benzene as a function of pressure and temperature; 273.15 Kelvin = 0 °C, 1 GPa = 10,000 bar

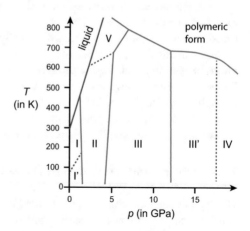

chemical laws that play a role in this issue are known. Crystallography shares this fate of knowing all the fundamental laws of nature but not being able to make reliable predictions with meteorology, which is not able to accurately predict the weather for more than 3 days. At least this unequivocally applies to the author's hometown (Hamburg, Germany).

The computer programs that are used to calculate crystal structures have made enormous progress over the last 20 years. Nevertheless, the old dream – "show me the chemical constitution formula and the shape of a single molecule, and I will tell you what crystal structure it will form" – remains unfulfilled yet. There is a simple reason why it is now possible to successfully reproduce the majority of polymorphs and their structures occurring in benzene with the aid of computer simulations [4]: benzene is a comparatively simple molecule; it has a disklike shape, is rigid, and has no flexible side chains. The calculation of crystal structures of complex molecules, with many conformational degrees of freedom, with competing possibilities of interaction between molecules, etc., is correspondingly difficult. The reason for this is usually that the energy values or stabilities of the crystal structures possible or conceivable under certain conditions do not differ sufficiently – the accuracy of the computer programs which calculate these energy values is still too low. In the case of very flexible molecules, it is also difficult to detect all structures of maximum stability at all. In recent years, however, astonishing progress has also been made here through the development of sophisticated algorithms – for example, genetic algorithms or Monte Carlo methods with simulated annealing. Since 1999, a kind of scientific competition has been held about every 2 years, in which research groups from all over the world have been invited to submit crystal structure proposals for a certain set of five to ten molecules selected by the CCDC, the organizer of this competition. The success rate has increased from competition to competition: in 2015 the prediction of possible polymorphs was also required for the first time. The team of Marcus Neumann of the Avant-garde Materials Simulation GmbH in Freiburg was able to correctly predict eight of the nine target structures – very remarkable [5]!

7.2.3 The Phenomenon of the "Lost" Polymorphs

Imagine the following scenario: you are the head of laboratory in a pharmaceutical company and initiate the synthesis of a new active substance. Your employees try out various syntheses to obtain the target molecule. The subsequent analyses prove that several paths led to success. Two different polymorphs of the same substance have been obtained. First medical studies have shown that polymorph I is more effective and has fewer side effects than polymorph II, so they hurry to the patent office and file a corresponding patent specification. Back in the laboratory, you instruct the synthesis of the new active substance on a large-volume scale. But now it turns out, although you made sure that the synthesis conditions were exactly the same, that either mixtures of polymorphs I and II are formed, in which the proportion of II clearly predominates, or only polymorph II is formed. You contact the head of laboratory of the foreign branch and delegate the original synthesis, which resulted in polymorph I, to be reproduced in the laboratory there. Curious how you are, you decide to get on a company jet and watch the action on site – impatiently you wait for the first results. These are devastating: even in the foreign branch, only the unwanted polymorph II is always obtained. Apparently, there is no trace of the polymorph I; any attempts to reproduce its synthesis fail. Polymorph I has "disappeared," which means more precisely "a crystal form that has been produced at least once and whose existence has been experimentally proven by any observation or measurement cannot be reproduced in subsequent attempts to produce the same crystal form by the same process; they lead to a different crystal form, alone or together with the old one" [6].

How can a polymorph disappear? The Gibbs' phase rule states that under exactly specified values of temperature and pressure, only one thermodynamically stable polymorph of a substance exists. Nevertheless, it is possible that several polymorphs are formed if, considered in absolute terms, the energetically less stable form is *metastable*, i.e., does not immediately transform into the more stable form, because this transformation is kinetically inhibited. However, once the more stable form has been formed, its presence can lead to an increased or even exclusive formation of this more stable form. This is an effect that could be described as unintentional inoculation. The inoculation, i.e., the addition of tiny seed crystals to a solution in which more and larger crystals are then to form, is a method also frequently used in pharmaceutical technology, in order to induce or considerably accelerate the crystallization from this solution. The hypothetical scenario described above can therefore be interpreted in such a way that after the first synthesis of the more stable polymorph II, it has contaminated the laboratories with finely distributed crystallization nuclei, so that now more or exclusively polymorph II is produced. The extent of this contamination may be difficult to imagine for laymen. But the reason that in the above story even in the remote laboratory only polymorph II is obtained could have actually been that the laboratory manager has acted as a carrier of the "wrong" seeds. The air contains an unbelievably high number of submicroscopic suspended or dust particles, to which crystal nuclei can be attached, either during the synthesis or the subsequent drying processes, etc., and thus can also be transported.

Does it mean that a "disappeared" polymorph can never be restored? The scientific community agrees that this is *not* the case. With reference to the Gibbs' phase rule, it is "only" to find the right temperature and pressure conditions. However, in practice this could mean to search for the famous needle in a haystack.

If you are interested in further background on this phenomenon, the author recommends a recent review article on "disappeared" polymorphs, which is an incredibly thrilling read. It comprises some older but also current cases of "disappeared" polymorphs in the pharmaceutical industry and also references to patent litigation [6].

7.3 Rock Salt: A Simply Complicated Structure and the Miracle of a Site with a Multiplicity of 192

In this section we will discuss some structural features of table salt or rock salt. Some of you may think that this is a rather boring substance being already sufficiently discussed at school. The author thinks that this is only half the truth, because from the crystallographic point of view, common salt or halite (the mineralogical name) is by no means a simple matter. In addition, due to the highly symmetrical arrangement of the atoms or ions, this structure reveals a certain kind of beauty.

Chemically viewed, common salt is an ionic compound consisting of single positively charged sodium ions and single negatively charged chloride ions, $NaCl$. It crystallizes in the space group $Fm\overline{3}m$ (space group number 225). From this, the crystal class $m\overline{3}m$ can be derived, which belongs to the cubic crystal system. For symmetry reasons, all angles must be set to 90°, and all lattice vectors must necessarily have the same length ($a = b = c = 5.625$ Å). The Bravais type is indicated with F, i.e., an all-sided face-centered lattice is present. And since our motif consists of two chemical species, both types of ions form a face-centered (partial) lattice, and the lattices are shifted in relation to each other by half of a unit cell (■ Fig. 7.18). Therefore, you have the choice to set either one of the sodium ions at the origin of the cell or one of the chloride ions.

If we look at the coordination environment of the ions, we realize that the sodium ions are octahedrally surrounded by six chloride ions and, inversely, the chloride ions are surrounded by six sodium ions in the form of a regular octahedron (■ Fig. 7.18). The coordination numbers are identical for both ions, since they each carry exactly one electrical charge, but the compound is electrically neutral overall.

The fact that sodium chloride is an ionic compound makes it relatively unlikely that you will find such large, well-developed halite crystals in nature as shown in ■ Fig. 7.19. Sodium chloride is very brittle and fragile. Although the attracting electrostatic forces between the positively charged sodium and negatively charged chloride ions are quite large, relatively

237   7

7.3 · Rock Salt: A Simply Complicated Structure and the Miracle of a Site ...

Fig. 7.18 The crystal structure of sodium chloride (space group $Fm\overline{3}m$), here shown as a $2 \times 2 \times 2$ supercell, in which two unit cells are shown in each of the directions x, y, and z. A single unit cell is marked in blue in the lower right corner. The octahedrally surrounded sodium ions (at the bottom left) and chloride ions (at the top right) are also highlighted; sodium is white, chloride green

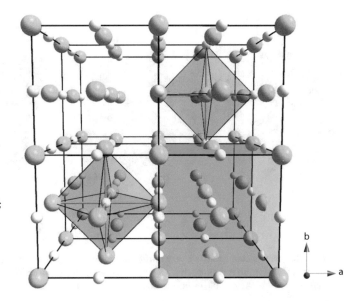

Fig. 7.19 Very large, well-formed, intergrown halite cubes (ca. $6.7 \times 1.9 \times 1.7$ cm) (Rob Lavinsky, ▶ iRocks.com, CC BY-SA-3.0)

small forces acting on one side of the salt cube can result in a situation, in which ions with identical signs are being directly adjacent. In order to realize this, the displacement has to be only one ion diameter (see ▣ Fig. 7.20). Since charges with identical sign repel each other, a crack forms at this point and finally the cube falls apart. In mineralogy, the term cleavage describes the property how easily a crystal can be cleaved. It also specifies in which

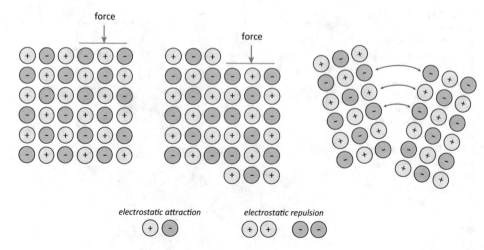

Fig. 7.20 Table salt or halite is a very brittle compound; since it is an ionic compound, a small force is sufficient to shift the ion rows in such a way that ion pairs of the same charge are directly adjacent, thus repelling each other. As a result, either cracks form or the material even divides into fragments

Cleavage and Fracture

In the world of mining and mineralogy, cleavage is defined as the property or tendency of minerals and rocks to break down into fragments along certain directions or planes that are structurally weaker than others under mechanical stress, for example, when struck with a hammer; the fragments have a certain number of smooth planes that depend on the specific crystal structure. The cleavage faces that are formed quite often coincide with the morphological growth faces of the minerals, i.e., they are mainly low-indexed faces, but they are independent of a certain initial form of the found object on which this mechanical stress is exerted.

Cleavage is one of the mechanical properties of minerals and is an important feature for identification that can contribute to the determination of minerals. Depending on the ease of cleavage and the quality of the formation of the cleavage surfaces, a qualitative distinction is made between:

- Highly perfect; many layered silicates of the mica group show this kind of cleavage, among them the single representative muscovite [$KAl_2[(OH,F)_2(AlSi_3O_{10})]$; and also gypsum (also called gypsum spar, $CaSO_4 \cdot 2 H_2O$) can be perfectly cleaved.
- Very good/perfect, as in the case of calcite ($CaCO_3$), fluorite (= fluorspar, CaF_2), baryte ($BaSO_4$), feldspars (including plagioclase feldspars), or galenite (= galena, PbS).
- Good, as, for instance, in the case of the calcium amphibole group (hornblende) or silicates of the pyroxene group.
- Distinct, for which olivine [$(Mg,Fe)_2SiO_4$] would be an example.
- Indistinct/poor, as in the case of corundum (Al_2O_3), or magnetite (Fe_3O_4).
- Absent, such as in quartz (SiO_2) or ilmenite ($FeTiO_3$).

You may have noticed that among the minerals with highly perfect or perfect cleavage, there are many spars. Since the early days of mining, spars have denoted minerals that show very good to perfect cleavability in several directions but whose cleavage surfaces show no metallic lustre; by the way, spar is just another word for parallelepiped.

In addition to the quality of the cleavage, a distinction is also made as to how many and along which directions minerals can be cleaved. This is exactly what is reflected in the crystal structure. Layer silicates have a distinct cleavage direction, namely, parallel to the layers along the directions {001}. The mineralogists would say "along the base pinakoid {001}." Some minerals are even named after their cleavage, e.g., orthoclase and plagioclase, whose cleavage surfaces form either a right ("ortho") or an oblique ("plagio") angle. For some minerals or mineral classes, specific angles (88° for pyroxenes, 124° for amphibole) are found. This characteristic can be used for the identification of these minerals. Common salt or halite (NaCl), sylvite (KCl), and galena (PbS) are examples of minerals that form cube-like cleavage products with three cleavage surfaces along the equivalent surfaces {001}. The trigonal carbonates magnesite ($MgCO_3$), calcite ($CaCO_3$), dolomite ($CaMg(CO_3)_2$), and siderite ($FeCO_3$), for example, also split along three cleavage surfaces, where rhombohedral cleavage bodies are preferred. Fluorite is a typical example of a mineral with four cleavage directions; it cleaves according to the octahedron along the four equivalent surfaces {111}. Six directions of cleavage (along {110}) can finally be found in the cubic sphalerite (zinc blende), which often forms well-developed rhombic dodecahedron as a cleavage body. In the table-like overview of ◘ Fig. 7.21, the information listed here is once again summarized according to the increasing number of cleavage directions.

A distinction must be made between cleavage and fracture. Fracture is also caused by mechanical stress on the material, but unlike in the case of fracture, the broken edges/faces do not correspond to specific surfaces that could be derived from the crystal structure. Some minerals, such as quartz or diamond, show only fracture under tensile or compressive stress, while other minerals may show both specific cleavages and irregular fractures. The resulting broken edges are in turn subdivided phenomenologically according to their appearance into:

- Conchoidal: typical, rotund grooves at the fracture point, resembling a semicircular shell, with rather smooth, curved surface, which also occur when glass is broken; e.g., in quartz (SiO_2) and flint ($SiO_2 \cdot n\,H_2O$).
- Hooky/hackly: small, sharp barbs are formed at the point of breakage; this occurs first and foremost for plastically deformable metals (gold, silver, copper).
- Earthy: very finely crumbly, dull, completely lacklustre fracture, e.g., chalk ($CaCO_3$) or limestone.
- Fibrous: the fracture looks like filaments or fibers, sometimes resembling aggregates of small hairs.
- Splintery: here, elongated splinters are formed, and sometimes small particles are being detached along the fracture.

Furthermore, fractures or broken edges can be uneven, flat (but rough), or smooth.

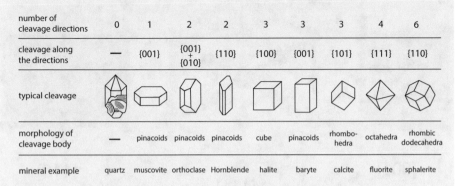

number of cleavage directions	0	1	2	2	3	3	3	4	6
cleavage along the directions	—	{001}	{001} + {010}	{110}	{100}	{001}	{101}	{111}	{110}
typical cleavage									
morphology of cleavage body	—	pinacoids	pinacoids	pinacoids	cube	pinacoids	rhombo-hedra	octahedra	rhombic dodecahedra
mineral example	quartz	muscovite	orthoclase	Hornblende	halite	baryte	calcite	fluorite	sphalerite

◘ **Fig. 7.21** Overview of cleavage directions and shapes of some typical minerals. (Redrawn, from: ▶ GeoDz.com ▶ http://www.geodz.com/deu/d/Spaltbarkeit)

direction the cleavage occurs. In the case of the sodium chloride, we find a "perfect cleavage" along all three equivalent faces {001}, i.e., parallel to the faces of the cube.

Sodium chloride is also a good example of how the crystallographic directions – which are slightly stressed in this book – can be brought to life and reproduced using the three-dimensional structure for the cubic crystal system. You can also identify the corresponding symmetry elements with the help of the VESTA or Mercury program. The long symbol of the space group 225 is $F\dfrac{4}{m}\bar{3}\dfrac{2}{m}$. This means that a four-fold axis of rotation and a mirror plane perpendicular to that axis must appear along the first viewing direction in the cubic crystal system (along [001]). If you rotate the structure in your preferred visualization software so that you view the structure along this viewing direction, both a four-fold axis of rotation and the mirror plane perpendicular to it (◻ Fig. 7.22a) should be recognized, although you must rotate the structure slightly around the viewing direction in order to clearly identify it.

If you orient the structure along one of the four space diagonals – it does not matter which concrete one you choose – you should see a similar picture as shown in ◻ Fig. 7.22b. At first glance it may look like something hexagonal, but when you look at the sodium ions drawn in gray, you will see that only a rotation of 120° leads to a congruent configuration of the structural components. The three-fold rotational symmetry along all four diagonals is the characteristic feature of the cubic crystal system. Furthermore, you should recognize the inversion center by slightly rotating the structure, because here not a simple three-fold axis of rotation is present, but a three-fold rotoinversion axis (" $\bar{3} = 3 + \bar{1}$ ").

To examine the crystal structure in the third viewing direction (along [110]), you must align it along the face diagonals. Now, it can be clearly seen that only a two-fold axis of rotation and a mirror plane perpendicular to it are present (◻ Fig. 7.22c).

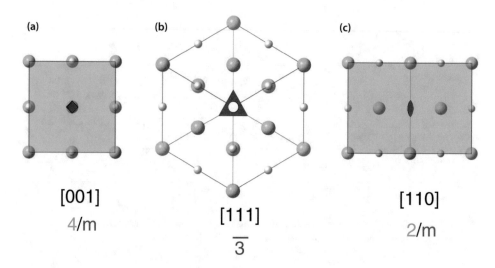

(a)   (b)   (c)

[001]
4/m

[111]
$\bar{3}$

[110]
2/m

◻ **Fig. 7.22** View of the crystal structure of NaCl along the three crystallographic directions: (a) [001]; (b) [111]; (c) [110]. The symmetry elements indicated in the long symbol of space group 225 are immediately recognizable

Fig. 7.23 The symmetry element diagram of the space group $Fm\bar{3}m$. (Reproduced with kind permission of the © International Union of Crystallography, ▶ http://it.iucr.org/)

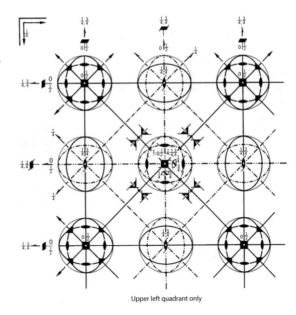

Upper left quadrant only

Up to now, the structure of sodium chloride seems rather straightforward and anything but complicated. However, if we take a look at the symmetry element diagram (**■ Fig. 7.23**) in the International Tables, we could come to the opposite conclusion. Please note that for the sake of clarity (and due to the existing symmetry), only the upper left quadrant is shown. This diagram looks relatively complicated and shows an abundance of symmetry elements, which have a corresponding duplication effect on objects at general positions.

Under the entry "Positions" in the International Tables, we can look up how many and which coordinates for the space group $Fm\bar{3}m$ are given for the general position (**■ Fig. 7.24**).

Here, it is noticeable that the multiplicity of the general position is indicated with 192, but only 48 coordinates are listed. However, above the entry for the coordinates, we find an additional line with the specification:

$$(0,0,0)+ (0, ½, ½)+ (½, 0, ½)+ (½, ½, 0)+$$

These are the specification for the centering. They have not occurred in the examples yet, because so far only primitive lattices have been discussed. In the present case of face centering, for each coordinate set that is explicitly listed under the coordinate entries and that belongs to the centering vector $(0,0,0)+$, there are three additional sets for which the fractional coordinates specified in parentheses must be added to each coordinate triple. This means, for example, that in addition to the coordinates

(1) x, y, z (2) $-x, -y, z$ (3) $-x, y, -z$ (4) $x, -y, -z$ etc.

also the coordinates

(1)' $x, y + ½, z + ½$ (2)' $-x, -y + ½, z + ½$ (3)' $-x, y + ½, -z + ½$
(4)' $x, -y + ½, -z + ½$, etc. and

◻ Fig. 7.24 The first entry in the section "Positions" for the space group $Fm\bar{3}m$. (Reproduced with kind permission of © International Union of Crystallography, ▶ http://it.iucr.org/)

Positions

Multiplicity, Wyckoff letter, Site symmetry

Coordinates

$(0,0,0)+$ $(0,\tfrac{1}{2},\tfrac{1}{2})+$ $(\tfrac{1}{2},0,\tfrac{1}{2})+$ $(\tfrac{1}{2},\tfrac{1}{2},0)+$

192 l 1

(1) x,y,z	(2) $\bar{x},\bar{y},z$	(3) $\bar{x},y,\bar{z}$	(4) $x,\bar{y},\bar{z}$
(5) z,x,y	(6) $z,\bar{x},\bar{y}$	(7) $\bar{z},\bar{x},y$	(8) $\bar{z},x,\bar{y}$
(9) y,z,x	(10) $\bar{y},z,\bar{x}$	(11) $y,\bar{z},\bar{x}$	(12) $\bar{y},\bar{z},x$
(13) $y,x,\bar{z}$	(14) $\bar{y},\bar{x},\bar{z}$	(15) $y,\bar{x},z$	(16) $\bar{y},x,z$
(17) $x,z,\bar{y}$	(18) $\bar{x},z,y$	(19) $\bar{x},\bar{z},\bar{y}$	(20) $x,\bar{z},y$
(21) $z,y,\bar{x}$	(22) $z,\bar{y},x$	(23) $\bar{z},y,x$	(24) $\bar{z},\bar{y},\bar{x}$
(25) $\bar{x},\bar{y},\bar{z}$	(26) $x,y,\bar{z}$	(27) $x,\bar{y},z$	(28) $\bar{x},y,z$
(29) $\bar{z},\bar{x},\bar{y}$	(30) $\bar{z},x,y$	(31) $z,x,\bar{y}$	(32) $z,\bar{x},y$
(33) $\bar{y},\bar{z},\bar{x}$	(34) $y,\bar{z},x$	(35) $\bar{y},z,x$	(36) $y,z,\bar{x}$
(37) $\bar{y},\bar{x},z$	(38) y,x,z	(39) $\bar{y},x,\bar{z}$	(40) $y,\bar{x},\bar{z}$
(41) $\bar{x},\bar{z},y$	(42) $x,\bar{z},\bar{y}$	(43) x,z,y	(44) $\bar{x},z,\bar{y}$
(45) $\bar{z},\bar{y},x$	(46) $\bar{z},y,\bar{x}$	(47) $z,\bar{y},\bar{x}$	(48) z,y,x

$(1)''$ $x + \tfrac{1}{2}, y, z + \tfrac{1}{2}$ $(2)''$ $-x + \tfrac{1}{2}, -y, -z + \tfrac{1}{2}$ $(3)''$ $-x + \tfrac{1}{2}, y, -z + \tfrac{1}{2}$ $(4)''$ $x + \tfrac{1}{2}, -y + \tfrac{1}{2}, -z$, etc. and

$(1)'''$ $x + \tfrac{1}{2}, y + \tfrac{1}{2}, z$ $(2)'''$ $-x + \tfrac{1}{2}, -y + \tfrac{1}{2}, -z$ $(3)'''$ $-x + \tfrac{1}{2}, y + \tfrac{1}{2}, -z$ $(4)'''$ $x + \tfrac{1}{2}, -y + \tfrac{1}{2}, -z$, etc.

exist.

So, the question is why the sodium chloride structure looks so simple and so tidy. And, of course, the answer can only be: this is due to the fact that both the sodium and chloride ions occupy highly special positions, namely, those of Wyckoff positions $4a$ and $4b$, each with the site symmetry $m\bar{3}m$ and the coordinates 0,0,0, and ½, ½, ½, respectively. The content of an unit cell therefore corresponds to the empirical formula Na_4Cl_4.

To gain a better understanding of a general position with a multiplicity of 192, it is recommended to add a random atom at a general position to the structure, for example, with the software VESTA. Open the corresponding CIF file with VESTA, and add an atom of your choice with the fractional coordinates $x = 0.08$, $y = 0.16$, and $z = 0.24$ (◻ Fig. 7.25). Then try other positions that are less general and observe the changes in the structure.

Perhaps, next time when you put salt on an insufficiently salted dish, you will also think of the inner beauty and in particular the neat arrangement of a grain of salt.

7.4 Graphite and Diamond

In this section we will look at the structure-property relationships of two of the three modifications of the element carbon: graphite and diamond. The third allotropic form of carbon, the C_{60} molecule fullerene, resembling a football, is very rare in nature, but is present to a certain amount in shungite coal (also known as algae coal). Allotropy is the term used to describe the phenomenon when one and the same chemical element occurs in the same physical state in different structural forms (Greek for "changing" or "in a different

◘ **Fig. 7.25** Resulting image after addition of another random atom at a general position to the crystal structure of NaCl

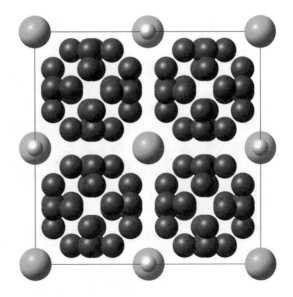

way"). The allotropic forms are also referred to as modifications of the element. The different structures lead to different physical and chemical properties.

7.4.1 Graphite

Graphite crystallizes in the space group $P6_3/mmc$ (space group number 194). It consists of individual, so-called graphene layers in which the carbon atoms form a hexagonal honeycomb structure. Within each of these layers, the carbon atoms form three chemical bonds to other atoms in the form of a regular triangle. These graphene layers, which lie in the (a,b) plane, are stacked perpendicularly to it, along the crystallographic c direction, with a layer distance of 335 pm. However, the individual layers do not lie exactly on top of each other, but are shifted relative to each other in the (a,b) plane. They form a so-called AB stacking sequence, i.e., every second layer is identical, or in other words, the third layer lies exactly above the first, the fourth exactly above the second, etc. (◘ Fig. 7.26).

If we look vertically at the graphene layers from above, we obtain an image according to ◘ Fig. 7.27. Every second atom of a layer is located in the center of six atoms of the underlying layer and vice versa. The dashed lines symbolize delocalized electrons; they are completely delocalized within a graphene layer, which explains the good electrical conductivity within the layers. This is parallel to the individual graph layers by a factor of 10,000 larger than perpendicular to them and reaches with a specific conductivity of $2.6 \times 10^4 \, \Omega^{-1} \, cm^{-1}$ almost metallic conductivity. Two adjacent layers, which are relatively far apart, are hold together not by real chemical (covalent) bonds, but only by relatively weak, so-called van

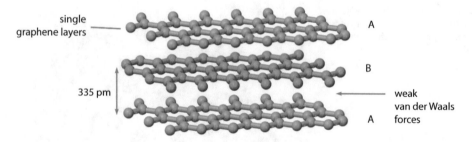
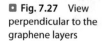

Fig. 7.26 Structural characteristics of the crystal structure of graphite, which consists of an alternating layer sequence (ABAB …) of individual graphene layers which are held together only by relatively weak van der Waals forces and have a distance of 335 pm (1 picometer = 1 billionth of a meter)

Fig. 7.27 View perpendicular to the graphene layers

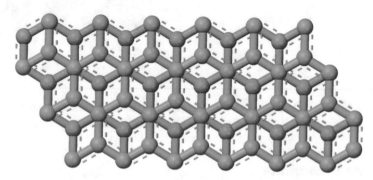

der Waals forces. This is also the reason why the overall thermal conductivity of graphite is relatively low.

With regard to the space group symbol, we see that there is a 6_3 screw axis along the first viewing direction and a mirror plane perpendicular to it. The first viewing direction in the hexagonal crystal system runs along the c direction. We can think about at which height along the c direction the mirror planes must be located. For this, we have to take into account the AB stacking sequence of graphite. The fact that two adjacent layers do not lie exactly on top of each other but are offset from each other in the (a,b) plane means that the mirror planes cannot be located between the graphene layers – they must, therefore, run through the graphene layers themselves. Along the second viewing direction, the crystallographic a direction, there is another mirror plane which cuts the hexagonal honeycombs in the middle or runs along their edges. And finally, along the third viewing direction, the direction [210], there is a glide plane c. This is oriented in such a way that the atoms of two adjacent layers are mapped onto each other (the translation component along the c direction is ½!).

The layered structure of graphite is also the reason for another property for which it is well-known: graphite is a very soft material, and it is used – together with clay – as a pencil lead. The explanation of the writing process lies in the friction that occurs when the pencil is pressed and moved on the paper, which ensures that successively individual graphene layers are being detached from the graphite bulk phase (also called delamination) – because they are only weakly bonded to one another – and are adhered to the paper (**Fig. 7.28**).

Fig. 7.28 The process of writing with a pencil, which is actually a graphite pencil, works because of the friction on the paper which causes individual graphene layers to peel off and adhere to the paper

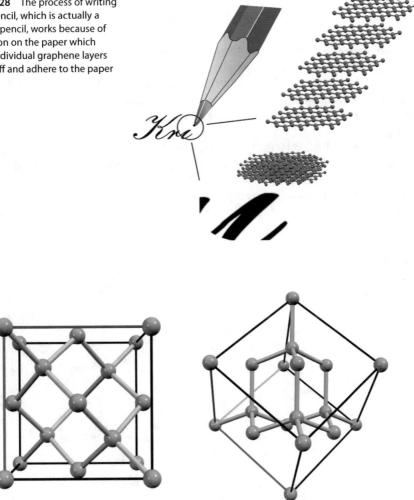

Fig. 7.29 Two views of the unit cell of diamond. Left, view along [001]; right, view approximately along [111]

7.4.2 Diamond

Let us now take a look at the crystal structure of diamond. **Figure 7.29** shows two views of the unit cell. The space group is $Fd\bar{3}m$ (space group number 227), which belongs to the cubic crystal system. Each carbon atom is bonded by four chemical bonds in the form of a tetrahedron to neighboring atoms; the chemical bonds run through the entire structure. Unlike graphite, there are no delocalized electrons. This network structure with strong, localized chemical bonds between all atoms explains why diamond is a bad electrical conductor but one of the materials with the best thermal conductivity. And it gives us an idea of why diamond – unlike graphite – is extremely hard. Diamond is the hardest mineral of all existing ones, with a value of 10 on the hardness scale according to Mohs. However, in recent times new synthetic materials have been produced that still have a somewhat higher hardness.

With regard to its symmetry, diamond can be recognized by the presence of a glide plane variant for which it is eponymous: the diamond-like glide plane. In diamond it runs perpendicular to the first viewing direction of the cubic crystal system (along [001]). If we visualize all existing glide planes with the help of the program Mercury, we see a confusing variety of planes that penetrate each other which is typical for the cubic crystal system (◘ Fig. 7.30).

Interestingly, diamond is the thermodynamically less stable form of carbon compared to graphite. And what we cannot recognize if we look at a diamond – and also not if we look at its crystal structure – is its high price. After all, it is not based on scientific reasoning but on the laws of the market economy (supply and demand). In the early Middle Ages, diamonds were not given a particularly high value. This gradually changed with the improved process-

7

◘ **Fig. 7.30**   Detail of the diamond structure; in addition, all diamond-like glide mirror planes are shown

Scale of Hardness According to Mohs
There are a number of different methods for assessing and determining the degrees of hardness of materials. There are both quantitative (absolute) degrees of hardness and qualitative (relative) degrees of hardness. A relative scale very common in mineralogy is the scale named after its inventor Mohs (Carl Friedrich Christian Mohs, German-Austrian mineralogist, 1773–1839). Mohs determined certain reference minerals for the hardness degrees of 1–10, the ranking of which was determined by scratching tests: if one mineral leaves a scratch on another when scribing, it is harder. The Mohs hardness scale is shown in ◘ Table 7.1.

If two minerals are directly compared and none can leave a scratch on the other, the two minerals are equally hard. If a mineral can scratch a certain mineral, but it cannot scratch the next harder mineral of the reference scale, its hardness lies between these two values. It is important to understand that the hardness scale according to Mohs is not a linear function; for better comparability of the actual hardness of the minerals mentioned, their hardness is also given in absolute terms, i.e. by the quantitative Vickers hardness (◘ Table 7.1).

◼ **Table 7.1** The Mohs hardness scale

Hardness	Mineral	Comment	Absolute hardness according to Vickers[a])
1	Talc ($Mg_3[Si_4O_{10}(OH)_2]$)	scrapable with the fingernail	2.4
2	Gypsum ($CaSO_4 \cdot 2\,H_2O$)	scratchable with the fingernail	36
3	Calcite ($CaCO_3$)	scratchable with a copper coin	109
4	Fluorite (CaF_2)	easily scratchable with a penknife	189
5	Apatite ($Ca_5[(F,Cl,OH)|(PO_4)_3]$)	with a penknife still just scratchable	536
6	Orthoclase ($KAlSi_3O_8$)	scratchable with a steel file	795
7	Quartz (SiO_2)	capable of scratching window glass	1,120
8	Topaz ($Al_2(F,OH)_2SiO_4$)	capable of scratching quartz	1,427
9	Corundum (Al_2O_3)	capable of scratching topaz	2,060
10	Diamond (C)	capable of scratching corundum	10,060

[a]In this process, a diamond pyramid is pressed with a fixed force on the material to be tested, and the indentation surface left by the pyramid in the material is determined using a microscope

ing capabilities of diamonds, particularly grinding and polishing, which made it possible to increase light reflection and refraction. The highlight of this cultural technique is the brilliant full cut, which was developed in 1910. This cut is able to bring off the "fire" of a gemstone.

7.5 Ferroelectrics: Isolators of a Special Kind and Modern Memory Components

In this last section of this chapter, we want to gain some insight into a very important class of materials for technical application: ferroelectrics. Tons of books have already been written about the physicochemical properties of this class of compounds and their material properties, and still new books are being added. The significance arises from the fact that ferroelectric materials are used on the one hand as capacitors and on the other hand find use in modern flash memory. Let us briefly consider the phenomenon of ferroelectricity and then take a closer look at the prototypical substance – barium titanate $BaTiO_3$ – in order to elucidate the structure-property relationships being responsible for the occurrence of ferroelectricity. Put in a nutshell, ferroelectrics are a special kind of dielectrics. Dielectrics are nonconductors or insulators that react to an electrical voltage with a polarization. In the case of ferroelectrics, however, two things are special: they are not only polarized in an electric field like all other

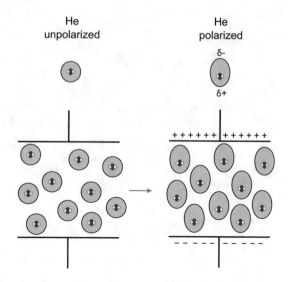

Fig. 7.31 Polarization effect on helium atoms in an electric field. This creates a small electric field inside in opposite direction to the external field (note that the charges in the helium atom are arranged exactly opposite to the external field)

dielectrics, but (a) the polarization also remains when the electric field is switched off, and (b) the direction of polarization can be inverted by applying an opposite electric field.

7.5.1 Displacement and Orientation Polarization

The term polarization describes the phenomenon that a substance forms a dipole in an electric field due to a spatial charge separation. However, there is no migration of charges in the electric field, as free charge carriers do not occur in insulators. This will become clearer later on. There are two types of polarization: the orientation polarization and the displacement polarization.

Displacement polarization occurs, for example, when a rare gas such as helium is introduced between the plates of a parallel-plate capacitor (■ Fig. 7.31). The electric field of the plates causes the negatively charged electron clouds to shift slightly to the positive pole, while the positively charged atomic nuclei slightly move to the negative pole. As a result, the center of charges of the negative and positive charge of the helium atoms is no longer identical – there is a slight spatial charge separation. Or to put it another way: the electric field has generated an intra-atomic dipole moment inside the helium atoms. However, this will only last as long as the external electric field remains switched on. Note that the totality of all dipoles of the helium atoms now generates an electric field itself, which is opposite to the outer one. The strength of this induced field can be measured simply by measuring the attenuation of the external electric field inside the parallel-plate capacitor. This attenuation factor (which is dimensionless) was formerly called dielectric constant (ε_r). Today it's called relative permittivity. The relative permittivity of helium is very small; it is only slightly above one, which means that the external electrical field is only slightly attenuated. The polarization effect is therefore only very weak which, however, is partly due to the fact that helium as a gas has a very low density and therefore there are only a few dipoles per volume element between the plates of the capacitor.

◻ Fig. 7.32 Water molecules that have a permanent dipole moment are being reoriented in an electrical field according to the direction of the field, so that the positive partial charges point to the negative pole and the negative partial charges point to the positive pole. This representation is simplified in the sense that the thermal motion of the water molecules is not taken into account

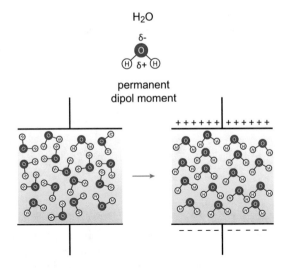

H_2O

$\delta-$

permanent dipol moment

◻ Fig. 7.33 The phase diagram of $BaTiO_3$; 273.15 Kelvin = 0 °C, 1 GPa = 10,000 bar. (Redrawn after: [7])

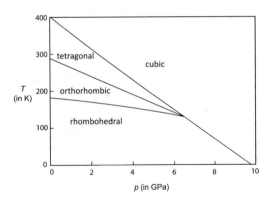

For substances that already have a dipole moment, such as water, displacement polarization also takes place, but the larger effect by far is the orientation polarization. It simply consists of the circumstance that the water molecules are oriented randomly before switching on an electric field but will align themselves uniformly according to the field lines after switching on the electric field (◻ Fig. 7.32). Water already has a relatively large field weakening potential, its relative permittivity is $\varepsilon_r = 79$.

7.5.2 $BaTiO_3$ and Spontaneous Polarization

Barium titanium oxide (also called barium titanate) is a solid that occurs in several modifications depending on the temperature (◻ Fig. 7.33) [7]. Above 120 °C (the melting point is 1620°), it is present in a cubic structure with the space group $Pm\overline{3}m$. In ◻ Fig. 7.34, a

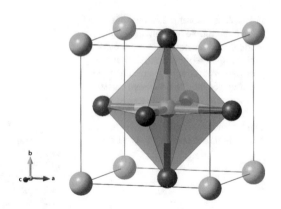

Fig. 7.34 The unit cell of BaTiO$_3$ with cubic structure; barium green, oxygen red, titanium light blue

unit cell is shown. At the center of the unit cell, a titanium cation is located being octahedrally surrounded by six oxide anions, which are located at the centers of the cell faces. The barium cations occupy the corners of the unit cell. This basic structure has its own name, it is the perovskite structure. Interestingly, the eponymous mineral perovskite (CaTiO$_3$) itself is not present in this prototypical structure type, but in an orthorhombically distorted phase. However, the related mineral tausonite, in which calcium is replaced by the heavier homologue strontium (SrTiO$_3$), is actually present in the perovskite structure. The cubic perovskite structure is characterized by the fact that the center of gravity of the negative (O^{2-}) and positive charges (Ba^{2+} and Ti^{4+}) is located at the center of the unit cell; therefore no dipole moment is present.

If BaTiO$_3$ is cooled below 120 °C, two very interesting things happen. First, a phase transition takes place from the so-called paraelectric cubic to the ferroelectric tetragonal phase (space group *P4mm*, number 99). This happens in such a way that the titanium ions move from the octahedral center in one of the six possible directions toward an oxide ion (■ Fig. 7.35). The displacement of the titanium species is only about 0.1 Å. Nevertheless, this tiny deflection is sufficient to induce a rather strong dipole moment (from the negative center of charge toward the titanium ion). This phase transition is called spontaneous polarization. Spontaneous because these dipoles originate quite quickly, if the material falls below 120 °C. What is actually astonishing, however, is the fact that the titanium ions are not individualists but show a remarkable collective behavior: in areas with diameters of several 10 to over 100 Å, the titanium ions move from the center of the octahedron in the *same* direction, corresponding to one of the possibilities shown in ■ Fig. 7.35. Therefore, a rectified (uniformly aligned) polarization occurs in these coherent areas, which are called domains. But overall, the material consisting of many such domains is initially unpolarized, since domains with all six possible polarization directions occur with equal probability.

7.5.3 Ferroelectricity: Permanent Polarization Without Field and Inversion of the Polarization Direction

How does ferroelectric barium titanium oxide behave in an electric field? With increasing field strength, the dipoles in more and more domains align themselves according to the direction of the electric field until at some point at very high external field strengths practically all domains have the same direction of polarization (■ Fig. 7.36a). The illustration

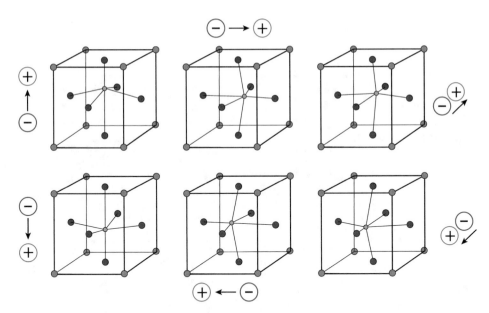

⬛ **Fig. 7.35** The six possibilities of spontaneous polarization of BaTiO$_3$ at temperatures below 120 °C. The titanium ion moves out of the center of the octahedron in the direction of one of the oxide ions. This induces a dipole moment

is somewhat simplified because it neglects the fact that domains with a favorable orientation of their dipoles grow at the expense of those that have an unfavorable, even opposite orientation to the external electric field.

If the electric field is now reduced and finally switched off completely (⬛ Fig. 7.36b), then the existing polarization remains essentially the same, only sporadically and due to thermal energy, the dipoles of some domains are reorienting. The strength of the remaining total polarization without an electric field is called remanence (lat. *remanens* = I remain). This phenomenon, known as ferroelectricity, is completely analogous to the related but much better known ferromagnetism. If a ferromagnetic material, e.g., iron, is introduced into a magnetic field, it itself gets magnetized and retains this magnetization almost completely, even if it is no longer located inside a magnetic field – a permanent magnet has been created.

In order to achieve a state in which a net polarization of the ferroelectric material is no longer present, an electric field with opposite sign of a certain strength must be applied (⬛ Fig. 7.36c). This strength is called coercive field strength (coercive from Latin: taming). The larger it is, the better the ferroelectric material retains its polarization. BaTiO$_3$ is one of the materials that requires high coercive field strengths to "erase" the polarization. If the outer field is increased even more strongly into the negative beyond the coercive point, the polarization increases again, this time in the opposite direction until saturation is reached again, and all domains are completely polarized in the opposite direction (⬛ Fig. 7.36c).

If the outer field is slowly reduced to zero again, a remanent polarization remains, which is as large as in the positive case, but now has an inverse sign. To delete this inverse polarization, a coercive field strength is also required here, now with a positive sign (⬛ Fig. 7.36d). If the field is further increased beyond the coercive field strength, the

7

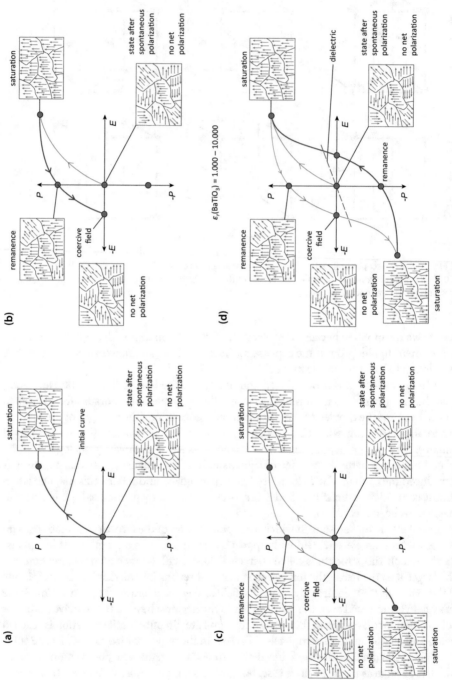

□ **Fig. 7.36** (a–d) Polarization of a ferroelectric material as a function of the external electric field (see text)

degree of polarization increases further until it has again reached the saturation value. Note that the very first initial curve cannot be reached again unless you heat the material above 120 °C, and then let it cool down again. Polarization as a function of electric field strength shows a typical hysteresis loop for ferroelectric materials (gr. *hysteros* = lagging behind) when a complete cycle is run through between two states of maximum saturation. Ordinary dielectrics show a completely different behavior, which is also shown in ◻ Fig. 7.36d for comparison: Here, the polarization is a linear function of the electric field strength, and no remanent polarization is maintained when the field is switched off.

7.5.4 Fields of Application: Capacitors and Memory Modules

The remanence behavior and the extremely high values of the permittivity of barium titanium oxide and other ceramics similar in structure (which have ε_r values of approx. 1000–10,000, depending on temperature and purity) make them excellent materials for storing (intermediate) electrical energy. Therefore, they have been used as capacitors in electrical circuits for quite some time (from approx. 1942 on). The use as nonvolatile memory module in FRAM modules (FRAM = ferroelectric RAM, RAM = random access memory) becomes more and more important. Probably, you have already guessed the operating principle: We have two distinguishable, defined states, namely, completely polarized in one or the opposite direction. These states can be defined as 0 and 1. These states remain even without a field, i.e., without an external voltage source. So, we have a memory module in front of us that doesn't need batteries. Only the following questions remain: How fast can the state be reprogrammed from 0 to 1, i.e., is the repolarization fast or slow? How much voltage or power is required for this? And can the switching process be repeated as often as required: what about the lifetime of such components? The answers: the switching process is extremely fast (150 nanoseconds), only a very low power is required and with 10 trillion read/write cycles the lifetime is very high. The FRAM modules based on ferroelectrics are thus clearly superior to conventional EEPROM and FLASH memories (e.g., the USB memory sticks are based on the latter). Only the SRAM modules (SRAM = static random access memory) – albeit volatile – outperform FRAM modules with regard to the speed of data transfer or writing speed (approx. 55 nanoseconds) and the lifetime, which is virtually unlimited.

7.5.5 Final Structural Consideration of BaTiO$_3$: Polar Axes – Pyro-Piezo-Ferro

The phase transition of the barium titanate from the paraelectric to the ferroelectric phase is associated with a symmetry break: the small but decisive displacement of the titanium ion from the center of the octahedron causes the loss of all three-fold axes of rotation, and instead of the three equivalent four-fold axes of rotation (parallel to the three axes of the coordinate system), there is only one, along the distinguished axis of the tetragonal crystal system, the *c* axis. This would also be the case if the titanium-oxygen octahedron was distorted by a symmetrical stretching or compression along one of the axes of the octahedron, which would lead to the space group P4/*mmm*. However, the displacement of the titanium ion from the equatorial plane of the octahedron results in a further symmetry break: the four-fold axis of rotation is still present parallel to the *c* direction, but it is characterized by the circumstance that perpendicular to it, no mirror

(a)   **(b)**   **(c)**

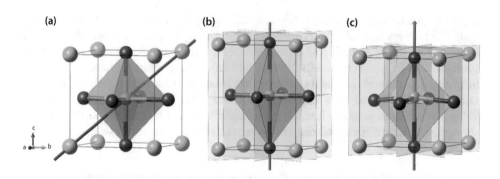

■ **Fig. 7.37**   Comparison of the (**a**) cubic phase of BaTiO3 (space group *Pm3̄m*) with a hypothetical, tetragonally distorted variant in which the TiO$_6$ octahedron is stretched only along the *c* axis (space group *P4/mmm*) (**b**) and the real, tetragonally distorted variant in which the titanium ion is moved out of the center of the octahedron (space group *P4mm*) (**c**). The cubic phase is characterized by four three-fold axes of rotation, one of which is marked as a red line. These are omitted in the tetragonal phases, which are characterized by a four-fold axis of rotation. In the ferroelectric phase, there is no mirror plane perpendicular to this four-fold axis, meaning that this is a polar axis

plane exists any more – compare ■ Fig. 7.37. Such axes are called *polar axes* because their presence in a crystal structure results in "polar" properties of the crystal, in the case of the tetragonal BaTiO$_3$ the formation of the electrical dipoles and the associated phenomenon of ferroelectricity.

There are two other phenomena related to ferroelectricity: piezoelectricity and pyroelectricity. Piezoelectricity (gr. *piezo* = "I press") refers to the phenomenon first discovered by Pierre Curie in 1880 and means that the prism surfaces of quartz crystals upon mechanical deformation are charged electrically positively and negatively, thus generating an electrical voltage. This can also be reversed: if a voltage is applied to a quartz crystal, it deforms. When the voltage is removed, the crystal relaxes, producing a vibration of characteristic duration – and thus forms the basis for the accuracy of quartz watches. The function of pick-up heads and speaker diaphragms is also based on the piezoelectric effect.

Pyroelectricity (gr. *pyros* = I burn), on the other hand, describes the property of some piezoelectric crystals to react to a temporal temperature change with a charge separation. The resulting voltage can be measured on the surfaces of the crystal. This effect is mainly used in sensor technology, e.g., for infrared detectors, motion detectors, or temperature sensors.

There are structural requirements for the presence of one or more of these three related properties. These are necessary but often insufficient. A classification can be made as follows:

1. Starting from all dielectrics, there is a subgroup that shows piezoelectric behavior. The necessary condition is that the crystal structure does not have a center of inversion ($\bar{1}$) or in other words that the associated crystallographic point group is not centrosymmetric. However, there is one exception of this rule: although crystals belonging to the cubic crystal class 432 do not have a center of inversion, they are *not* piezoelectric.

2. Of all piezoelectric compounds, there is a smaller subgroup, which is also pyroelectric. The necessary prerequisite here is the presence of at least one polar axis.

☐ Fig. 7.38 The subsets of all dielectrics

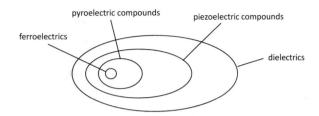

3. Among the pyroelectric materials, there is an even smaller subgroup showing ferroelectric behavior. For ferroelectrics, the necessary prerequisites are those for both piezoelectric and pyroelectric compounds.

The classes of compounds that are subsets of other classes of compounds are graphically summarized in ☐ Fig. 7.38.

References

1. Mercury 4.1.0. The Cambridge Crystallographic Data Centre: https://www.ccdc.cam.ac.uk/Community/freeservices/freemercury/ (2019) Accessed 30 Sept 2019
2. Desiraju GR (2007) On the presence of multiple molecules in the crystal asymmetric unit (Z' > 1). Cryst Eng Comm 9:91–92. https://doi.org/10.1039/b614933b
3. Steiner T (2000) Frequency of Z' values in organic and organometallic crystal structures. Acta Cryst B 56:673–676. https://doi.org/10.1107/S0108768100002652
4. Raiteri P, Martoňák R, Parrinello M (2005) Exploring polymorphism: the case of benzene. Angew Chem Int Ed 44:3769–3773. https://doi.org/10.1002/anie.200462760
5. Gibney E (2015) Software predicts slew of fiendish crystal structures. Nature 527:20–21. https://doi.org/10.1038/527020a
6. Bučar D-K, Lancaster RW, Bernstein J (2015) Disappearing polymorphs revisited. Angew Chem Int Ed 2015 54:6972–6993. https://doi.org/10.1002/anie.201410356
7. Hayward SA, Salje EHK (2002) The pressure–temperature phase diagram of BaTiO$_3$: a macroscopic description of the low-temperature behavior. J Phys Condens Matter 14:L599–L604. https://doi.org/10.1088/0953-8984/14/36/101

"Forbidden" Symmetry

© Springer Nature Switzerland AG 2020
F. Hoffmann, *Introduction to Crystallography*, https://doi.org/10.1007/978-3-030-35110-6_8

Until 1982 it was considered an irrefutable paradigm in crystallography that crystals cannot have rotational symmetry of the order of five or higher than six. This limitation of rotational symmetry has already been mentioned several times in this book (see ► Sects. 3.2 and 5.2). This limitation applies to both the order of axes of rotation as well as screw axes: only axes with the order 1–4 and 6 occur in each case. However, we will discover in ► Sect. 8.4 that this is not the whole truth.

8

8.1 Impossible Lattices

But first we want to shed some light on this paradigm by showing that orders other than 1–4 and 6 cannot be reconciled with a strictly periodic lattice – such rotational symmetries are incompatible with the periodicity of crystals. We can verify this, for example, by trying to construct a lattice with five-fold rotational symmetry. In ◘ Fig. 8.1a a section of a centered 2D lattice with five-fold rotational symmetry is shown, with the blue points marking the lattice points and the lattice vectors $\vec{a}$ and $\vec{b}$ running along the edges of the pentagon. If these were actually lattice vectors, then a translation of the lattice along these vectors should lead to a congruent image. As shown in ◘ Fig. 8.1b, this is not the case; a shift along the lattice vector $\vec{b}$ would result in an analogue image. Perhaps the lattice vectors have been incorrectly determined and a congruent image is obtained by moving along the vector $\vec{a}'$ in ◘ Fig. 8.1c? ◘ Figure 8.1d shows that this is not the case either. There are simply no vectors at all that would meet the definition of lattice vectors for a pattern with five-fold rotational symmetry – in other words, there is no lattice. Just as in this example with five-fold symmetry, you will not be able to construct periodic lattices with a rotational symmetry of the order larger than six. Note, however, that it is

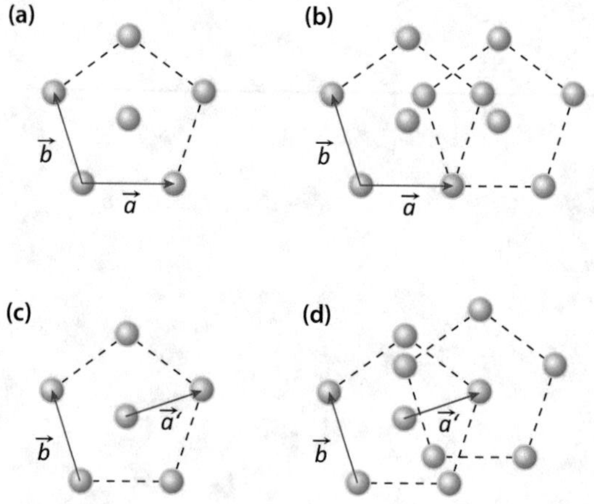

◘ Fig. 8.1 (a–d) Checking whether there can be a lattice with five-fold rotational symmetry shows that there are no lattice vectors that map the lattice points onto each other – ergo: the point formations shown here do not form a lattice

of course possible to form a lattice in which the motif has a five-fold rotational symmetry (see ◘ Fig. 8.2).

8.2 Possible and Impossible Tessellations

We can also illustrate the incompatibility of lattices with bodies that have a five-fold, seven-fold, and higher order of rotational symmetries geometrically in such a way that we try to fill the plane without gaps with tiles of corresponding rotational symmetry, i.e., to create a so-called tessellation.[1] As shown in ◘ Fig. 8.3, this is actually only possible for polygons with an order of rotational symmetries of one to four and six. In all other cases, gaps remain.

◘ **Fig. 8.2** An arrangement of a motif with five-fold rotational symmetry in an oblique-angled lattice

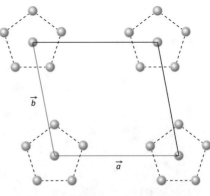

◘ **Fig. 8.3** When attempting to completely cover the plane with polygons with five-fold or larger than six-fold rotational symmetry (without overlapping), gaps always remain

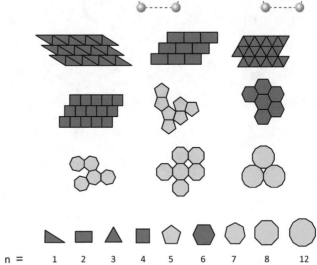

---

1   This term is equivalent to the term tiling and describes in mathematics the gapless and overlap-free covering of the (Euclidean) plane by uniform sub-areas.

Tessellations by Pentagons

Although it is impossible to cover the plane without any gaps and without overlap with a regular pentagon, this is possible with non-regular pentagons. But how do they look like, and is there an instruction on how to use the pentagons to construct a tessellation? And are there an infinite number of those irregular pentagons that cover the plane? These questions have been addressed in mathematics for about 100 years. The hunt for convex pentagons (i.e., those in which all internal angles are less than 180°), which allow complete covering of the plane, began with the discovery of five classes of pentagons by the German mathematician Karl Reinhardt in 1918 [1]. Classes mean that these are not five individual pentagons but five types, each of which can be described by a system of equations for the relative edge lengths and angles, whereby the concrete appearance of a pentagon of one class can be slightly different. Three more classes were discovered by physicist Richard B. Kershner of Johns Hopkins University in 1968 [2] and another class 7 years later, in 1975, by computer scientist Richard E. James. His discovery was published in *Scientific American* [3] and was read with great interest by Marjorie Rice, who was passionate about puzzles but otherwise had no higher mathematical education. For Rice it was an exciting

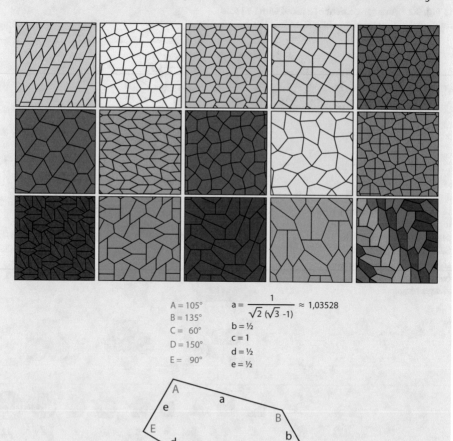

$$A = 105°$$
$$B = 135°$$
$$C = 60°$$
$$D = 150°$$
$$E = 90°$$

$$a = \frac{1}{\sqrt{2}\,(\sqrt{3}-1)} \approx 1{,}03528$$
$$b = \tfrac{1}{2}$$
$$c = 1$$
$$d = \tfrac{1}{2}$$
$$e = \tfrac{1}{2}$$

□ **Fig. 8.4** All 15 classes of tessellations by non-regular, convex pentagons. The recently discovered class is located in the last line on the far right and consists of a pentagon type specified with the given aspect ratios and angles

new puzzle to find out if there were more tessellations with pentagons. She began sketching corresponding diagrams, and just a few weeks later, she had discovered a new class of pentagons with tessellation properties. Over the next 2 years, she found two more. The number of pentagons tiling the plane had risen to 13. Another 7 years later, in 1985, Rolf Stein found the class No. 14 [4], but then it became quiet for a long time around the pentagons with the special characteristics. It was not until 30 years later, in 2015, that another one was found, the 15th class, this time supported by a computer program, which systematically varied the lengths of the edges and the angles between them [5]. In ◻ Fig. 8.4 all previously discovered 15 classes of tiling of the plane by non-regular pentagons are shown.

Does this mean that all classes have been discovered by now? We do not know. There is no proof that there are not any further classes of pentagons. It may sound strange, but the pentagon as such and its tiling properties are not very well studied. Now you might think that there are probably also tilings with irregular heptagons or octagons. Remarkably, this is not the case.

8.3 Possible and Impossible Morphologies

The limitation of rotational symmetry also applies to the morphology of crystals. There are no crystals whose outer forms have a five-fold, seven-fold, or rotational symmetry with higher order. This holds true even for crystals, which do not infrequently crystallize in such forms, which, both when looking at the shape and at its name, might suggest that there is a five-fold rotational symmetry, for example, pyrite, which may be present in the form of a pentagonal dodecahedron (◻ Fig. 8.5, see also ▸ Sect. 2.1).

Usually a pentagonal dodecahedron is a regular polyhedron with 12 faces (Greek *dodecahedron* = having twelve faces) in the form of congruent and regular pentagons as boundary faces. Such a *regular* pentagonal dodecahedron now also has five-fold rotational symmetry, the six existing five-fold rotational axes each run through opposite face centers. Therefore, the pyritic dodecahedron form cannot be a regular dodecahedron. In fact, it is a cubic pentagonal dodecahedron – matching the cubic crystal system in which pyrite crystallizes – which is therefore also called pyrite dodecahedron or pyritohedron. In this form of a pentagonal dodecahedron, the pentagons are no longer regular: one side of the pentagons is shorter, four are longer.

◻ **Fig. 8.5** Some selected shapes of pyrite crystals. In total more than 60 different shapes are known. The pentagonal dodecahedron is not a polyhedron with five-fold rotational symmetry (see text)

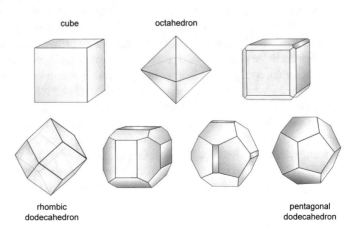

cube

octahedron

rhombic dodecahedron

pentagonal dodecahedron

8.4 Discovery of Quasicrystals: An Educational Act of History of Science

An extremely interesting discovery from 1982 posed an enormous challenge to crystallography: the discovery of quasicrystals by Daniel ("Dan") Shechtman. These are crystals with "forbidden" symmetry, involving 5-, 10-, and 12-fold rotational symmetries. The discovery and slow realization in the crystallographic community that these entities are a new, or rather hitherto undiscovered form of matter that has features of crystals but is not actually periodically ordered, is one of the most exciting episodes in the recent history of science, which will be briefly told below.

It is the year 1982: Daniel Shechtman (∗ 1951, Tel Aviv) – actually based at the Technion (Israel Institute of Technology) in Haifa, Israel – works as a visiting scientist at Johns Hopkins University in Baltimore (USA) on the development of novel aluminum-based alloys that may be used in the aviation industry. He investigated the metallic samples he obtained, among others with a technique called electron diffraction. This technique was still relatively young at that time, but the physical principles are the same as the long-established X-ray diffraction. This investigation method for crystalline samples shall not be explained in detail here, but its basic features are briefly described, since they are necessary for further understanding.

The use of X-rays in medical care is well known to everyone, but in this case it is a single-crystal rather than a broken leg that is radiated with X-rays. Here, the X-ray beam is not only scattered more or less diffusely – depending on the density of the tissue – but it is deflected in a specific way; it is *diffracted*. As the X-rays pass through, the rays interact specifically with the crystal, resulting in destructive (intensity attenuating) and constructive (intensity increasing) interferences depending on the structure of the crystal. If the X-rays are detected after the passage of the crystal with a photosensitive layer behind the sample, diffraction points or reflections at only very few locations can be seen, which result in a specific pattern (see ◘ Fig. 8.6). The symmetry of the pattern and the intensity of these X-ray reflections can be used to determine the atomic structure of the crystal. This technique is called X-ray crystal structure analysis. And in principle, electron diffraction works just like X-ray diffraction, except that here electrons are fired to the crystal sample.

The important thing for understanding Dan Shechtman's experimental results is that the symmetry of the examined crystal is reflected in the symmetry of the diffraction points. For example, if the examined crystal has hexagonal symmetry, then the diffraction points also form a pattern with hexagonal, i.e., six-fold rotational symmetry. Surprisingly, some of the samples examined by Shechtman showed diffraction patterns that no human being had ever seen before: those with 5-, 10-, or even 12-fold rotational symmetry (◘ Fig. 8.7)!

Shechtman could hardly believe his eyes! A diffraction image with ten-fold symmetry? Skeptically he wrote in his laboratory journal: "10 Fold???." Should he have found a new form of matter? With "forbidden" symmetry? How could that be possible? He was convinced he had something new in front of him, but for a long time he remained the only one. In retrospect that may be understandable: stating that crystals were created with "forbidden" symmetry would be like claiming in astronomy that a flat-disk planet was discovered. Shechtman had to struggle with disbelief and even malice. His director of research strongly recommended him to review the relevant findings from crystallography text-

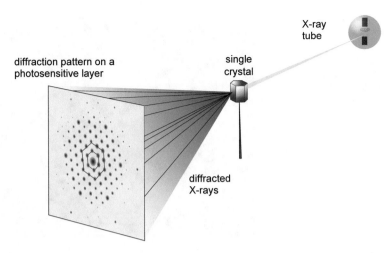

◘ Fig. 8.6 Principle of an X-ray diffraction experiment: coherent (in-phase) X-rays hit a single-crystal and are diffracted in a specific way as they pass through it. The diffracted X-rays are detected with a photosensitive layer. The position or pattern of the diffraction points is indicative of the symmetry of the crystal (here: hexagonal); the intensity distribution indicates the structure of the crystal

◘ Fig. 8.7 A diffraction pattern with ten-fold rotational symmetry obtained by irradiating a single crystal of a Zn-Mg-Ho alloy with electrons. Dan Shechtman's aluminum alloys produced analogous patterns. In addition to the ten-fold rotational symmetry (highlighted in blue), patterns with locally pentagonal symmetry can also be clearly seen (highlighted in orange). (CC BY-SA 3.0 Wikipedia User Materialscientist)

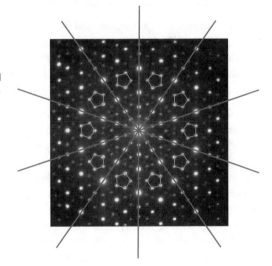

books. Many of Shechtman's fellow researchers assumed he made experimental errors or considered the possibility that he had not measured single-crystals but twins or multiples, i.e., pieces of crystal that look as if they were made of one single piece but in fact consist of two or more domains that have grown together. Some minerals are known to occur more frequently in a specific twin form, which results in the outer form having five-fold rotational symmetry – eponymous, for example, pentagonite (◘ Fig. 8.8), a deep blue, vanadium-containing silicate of the composition $Ca(VO)Si_4O_{10} \cdot 4\,H_2O$.

◘ Fig. 8.8 Pentagonite crystals that typically occur in the form of tufts. Individual aggregates from two domains, i.e., twins, sometimes appear to have a five-fold rotational symmetry. (CC BY-SA 3.0 Wikipedia User Hectonichus)

But Shechtman was a good experimentalist, and he trusted his results. He could definitely rule out twin formation. It took another 2 years for his experiments and conclusions to be accepted for publication in the prestigious scientific journal *Physical Review Letters*; the work is entitled "Metallic phase with long-range orientational order and no translational symmetry" [6]. The summary reads like this:

> ❯❯ We have observed a metallic solid (Al-14-at.%-Mn) with long-range orientational order, but with icosahedral point group symmetry, which is inconsistent with lattice translations. Its diffraction spots are as sharp as those of crystals but cannot be indexed to any Bravais lattice. The solid is metastable and forms from the melt by a first-order transition.

This was the birth of the quasicrystals – crystals with long-range orientational but no translational symmetry. On the one hand, these solid bodies lack the strict periodicity that characterizes a crystal – they are aperiodic – and on the other hand, in diffraction experiments with X-rays or electrons, they behave exactly like crystals. Only reluctantly, and especially through the help of the theoretical side, by Paul Joseph Steinhardt, who became aware of Shechtman's work, the realization finally prevailed that Shechtman had indeed discovered a new form of quasicrystalline matter and was not mistaken. Steinhardt also worked as a theoretical physicist on the question of whether there could be a form of quasiregular matter that led to icosahedral diffraction patterns, stimulated by the so-called Penrose tilings (▶ Sect. 8.5), and preparatory work by Alan Lindsay Mackay (∗ 06.09.1926, British physicist), who irradiated such Penrose patterns with light and also received diffraction patterns with icosahedral symmetry [7]. When he became aware of Shechtman's work by a preprint of the original article, he was immediately fascinated – Shechtman's diffraction patterns had to originate from compounds he was looking for! Only 5 weeks later, Steinhardt and his doctoral student Dov Levine published a paper on the theory of the occurrence of such "impossible" diffraction patterns, and they also coined the term quasicrystal [8]. In 2011 Shechtman (◘ Fig. 8.9 shows a portrait) was awarded the Nobel Prize for Chemistry for the discovery of quasicrystals [9].

■ **Fig. 8.9** Dan Shechtman with a wireframe model of an icosahedron, which constitutes one of the two types of quasicrystals, at the press conference for his Nobel Prize 2011. (Technion – Israel Institute of Technology; CC BA-SA 3.0)

Interestingly, there was one very famous exception among scientists who denied the existence of quasicrystals: no less than the famous chemist Linus Pauling himself, a double Nobel Prize winner (chemistry and peace), strictly rejected the existence of quasicrystals until his death in 1994. He is quoted as saying, "There are no quasicrystals, only quasi-scientists."

Despite intensive research on quasicrystals (there are now more than 10,000 publications on this topic), fundamental questions remain unanswered, for instance, the mechanism of formation or the exact atomic structure. A good overview of the current status until 2017 can be found in Walter Steurer's review article in Ref. [10]. Shechtman's discoveries revolutionized the ideas and views of classical crystallography. However, this was only due to Shechtman's considerable stamina. In an interview worth seeing or hearing (available on YouTube), Shechtman says when asked what his message is based on his experiences of the discovery of quasicrystals [11]:

» Hey, if you are a scientist and believing your results, then fight for them, then fight for the truth.

For those who are further interested in the processes of scientific revolutions in general, the author recommends the highly exciting book by the philosopher of science Thomas S. Kuhn: *The Structure of Scientific Revolutions* [12].

8.4.1 Natural Quasicrystals?!

Dan Shechtman dealt with artificial alloys when he discovered his quasicrystals. He obtained them by cooling the corresponding metal mixtures from the melt very quickly. This raises the question whether the occurrence of matter in quasicrystalline form is limited to such artificial alloys or whether there might not also be conditions in nature that at least allow the formation of quasicrystals. This question was investigated, among others, by Paul Steinhardt, who began to investigate the X-ray diffraction data sets of materials and minerals, looking for anomalies of quasicrystals. He also called on all colleagues in the world who are directors of mineralogical collections to do the same. And indeed, in 2009 Luca Bindi from the University of Florence came up with a hot candidate. It turned out that this candidate was eventually indeed the first confirmed quasicrystal of natural origin [13]. It was a rock specimen from the East Russian region of the Koryak mountains containing some pieces of metal that turned out to be perfectly quasicrystalline: an alloy of aluminum, copper, and iron. In further investigations and expeditions, it was also found that this rock sample came from the Kathyrka meteorite, which had descended in the Koryak mountains and has an estimated age of 4.5 billion years. The accompanying minerals of the meteorite suggested that the meteorite – shock-induced by a collision with another celestial body – was exposed to very high pressure, which resulted in high temperatures [14]. Apparently, it was just as quickly cooled again – thus ideal conditions for the formation of quasicrystals.

8.4.2 Crystals: A Redefinition

One question remains to be answered: Are quasicrystals crystals or not? On the one hand, they lack the characteristic par excellence which distinguishes crystals, namely, to exhibit a strict periodicity or a strict translational order. How could they be crystals? On the other hand, in a diffraction experiment, they generate defined patterns – and only crystals lead to such patterns with defined, delimited diffraction points! The International Union of Crystallography has decided to focus on the latter side and has reworded its definition of what a crystal is:

> A crystal is a solid form of matter that leads to defined Bragg peaks (i.e., diffraction points) in a diffraction experiment.

8.5 Penrose Tilings

The recognition of quasicrystals by the scientific community as a possible form of solid matter relied to a large extent on theoretical preparatory work in the field of mathematics. It has been shown above that it is not possible to completely cover the plane with regular pentagons; necessarily gaps remain. But as early as 1974, Sir Roger Penrose (∗ 08.08.1931, Colchester, English mathematician and theoretical physicist) showed that it is possible to design a tiling consisting of two different rhombi (a thin and a thick rhombus) which (a) is aperiodic and (b) has a (local) five-fold rotational symmetry [15]! In ■ Fig. 8.10a such a Penrose tiling is shown. Until that time, it was not believed that it is possible to completely cover the plane with something that is nonperiodic. It is not possible to define a section of the Penrose tiling that fulfills the definition of a unit cell and would result in the total covering of the plane by simply joining them together. If we take a closer look at the pattern, we discover different areas that look very similar, but they are not the same (■ Fig. 8.10b). This pattern has – fascinatingly – simply no recurring motifs.

However, when designing a nonperiodic tessellation, it is not sufficient to know how the building blocks have to look like, in the present case, which particular rhombic shapes one must put together. What is additionally required is the construction plan, because there are almost infinite possibilities to completely cover the plane with these two rhombi, and in most of the cases, something periodical will form! This means that the *aperiodic* tiling has been the actually remarkable achievement of Penrose.

It could also be shown that it is possible to form three-dimensional analogues to the 2D Penrose tilings, in which no rhombi, but rhombic dodecahedra, can be put together to completely fill the space without resulting in a periodicity. These rhomboid-like packages also form the basis of the atomic structure of quasicrystals, which elude a common description in 3D space. However, amazing progress has now been made in the structure elucidation of quasicrystals, so the question, "Where are the atoms actually located in a quasicrystal, if they are constituents of entities that defy description as (ordinary) crystals?" for some of these systems can be answered very well [16].

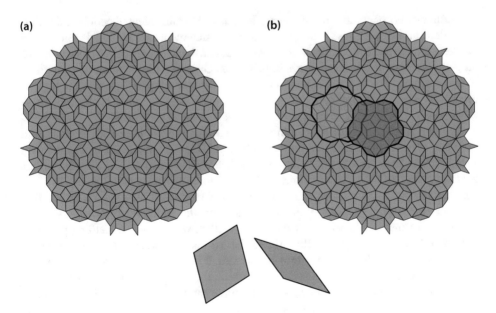

■ **Fig. 8.10** (a) An aperiodic Penrose tiling with five-fold rotational symmetry. It consists of the two rhomboids, which have to be laid according to a specific construction plan. (b) Two highlighted areas which look the same when viewed briefly are in fact not identical. There is no delimited area that ever repeats itself in this pattern

Quasicrystals cannot be described as regular or periodic structures in 3D space, but they can be described in six-dimensional space, which leads us to the following section "Crystallography in Higher Dimensions."

8.6 Crystallography in Higher Dimensions

When you start dealing with quasicrystals, you can quickly get dizzy. You're entering dimensions you can't really imagine. It is well known that mathematically gifted people often have fewer problems with it. The mathematics of higher (>3) dimensions is now very well developed. And therefore quasicrystals – apart from the fact that they exist and have interesting properties – are ideal entities to do mathematics with or on them, specifically mathematics of hyperspace.

There is a geometric interpretation of quasicrystals, which can be summarized in the following statements:

1. Any quasiperiodic pattern of points can be formed from a periodic pattern of a higher dimension.
2. To create a 3D quasicrystal, one can begin with a periodic array of points in a 6D space. The three-dimensional space should be a linear subspace that penetrates the six-dimensional space at a certain angle.
3. If you now project each point of the 6D space that is within a certain distance of the 3D subspace onto the subspace and the angle represents an irrational number, such as the golden ratio, then a quasiperiodic pattern is created!

The author does not count himself among the mathematically talented people and is thus dependent on help, because – to be frank – these statements are all Greek to him. An excellent aid are projections: breaking down to dimensions that are easier to survey. This have already been used several times in this book. Once one has understood something in lower dimensions to some extent, one is often content with the fact that the higher-dimensional case is "completely analogous." Let's give it a try. So, what is meant by these statements?

We start from a periodic pattern in two dimensions, a point lattice of the plane, and use a projection to make it an aperiodic structure in 1D space, i.e., a point arrangement on a straight line. In ◘ Fig. 8.11a an ordinary square Bravais lattice is shown. Now the points are to be projected onto a straight line, i.e., they are to be projected onto this line at a certain angle by pulling them down vertically onto the straight line. Now we can make a first consideration of the angle that the straight line with the lattice vectors encloses. If the straight line runs parallel to one of the lattice vectors, we only get something periodic, even exactly the periodicity of the original lattice (as indicated by the red arrows in ◘ Fig. 8.11a). The angle must, therefore, not be zero. It is also easy to see, for example, that it must not be 45°. According to the third statement, the angle should represent an irrational number, i.e., a decimal number with an infinite number of decimal places, which is nonperiodic! This can also be expressed in such a way that, for example, if the line starts at point (0,0), it must not intersect any other lattice point. As an opposite example, in ◘ Fig. 8.11b the line is chosen so that it hits the lattice point (5,2): Something periodic will result in this case, since the range between the lattice point (5,2) and the next lattice point

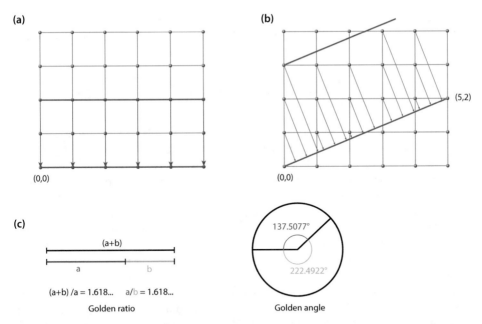

(a)

(0,0)

(b)

(5,2)

(0,0)

(c)

(a+b)

a   b

(a+b) /a = 1.618...   a/b = 1.618...

Golden ratio

137.5077°

222.4922°

Golden angle

☐ **Fig. 8.11**   (a) The projection of lattice points onto a straight line which runs parallel to one of the lattice vectors results in something periodic; (b) likewise, the angle of the projection of the straight line may not be selected in such a way that it intersects another lattice point, as here the point (5,2), because also then necessarily a periodicity is generated. (c) If, on the other hand, the angle is chosen in the ratio of the golden section (or in the ratio of another irrational number), the projection method along a straight line generates a one-dimensional quasicrystal

that intersects the line, lattice point (10,4), must be identical to the range between (0,0) and (5,2). It is also immediately obvious that the area in which this projection method is applied must not be infinitely large, as this would lead to an infinite number of points on the straight line. Exactly such an utterance is contained in statement 3, in which it is said that not all points are to be projected, but only those that lie within a certain, finite distance from the projection line. Hence, provided that the angle is chosen according to the ratio of the golden section (see below) and the area in which this projection method is applied is limited, something aperiodic, a one-dimensional quasicrystal is actually created! And "just in this way," through such a special projection, an aperiodic crystal in our known three-dimensional space arises out of the six-dimensional space in which this quasicrystal would be periodic.

It remains to be clarified what the golden ratio is, abbreviated with the large Greek letter Φ (Phi). The golden section is a certain ratio of two quantities, e.g., the division ratio of a section, where the ratio of the total length to the larger section corresponds to the ratio of the larger section to the smaller section. If this is the case, the ratio is

$$1 : \frac{1+\sqrt{5}}{2} = 1.618034 \dots$$ (☐ Fig. 8.11c). Remarkably, proportions that behave according to

the golden ratio are perceived by humans as particularly aesthetic and balanced, which is why the golden ratio is of paramount importance – not only in mathematics but also, for example, in art and architecture. This special ratio can be related not only to distances but also to angles. If we assume a full circle of 360°, then the golden ratio divides the circle into two parts: a larger angle of 222.4922... ° and a smaller one of 137.5077... °, because it applies:

$$\text{Golden Angle}: \frac{360^{\circ}}{222.4922\ldots^{\circ}} = 1.618034\ldots \quad \text{and} \quad \frac{222.4922\ldots^{\circ}}{137.5077\ldots^{\circ}} = 1.618034\ldots$$

This golden ratio is also found in nature, e.g., in certain plants where the leaves are arranged along a circle and the angle between two consecutive leaves divides the full circle in the ratio of the golden ratio, i.e., forms an angle of about 137.5° (◻ Fig. 8.12). Since the positions along the circle are never repeated, the leaves have a maximum total surface area that can be reached by the sunlight, so that effectively photosynthesis can be carried out. The petals of the sunflower are an example for this kind of arrangement.

The golden ratio is found at quasicrystals not only in the form of the abovementioned projection relationship or as an interface between hyper space and 3D space but also in the form of its external shape. Single-crystalline quasicrystals have the form of either a regular (!) pentagonal dodecahedron or an icosahedron. At both polyhedra you will find the golden ratio. The faces of a pentagonal dodecahedron are formed by regular pentagons. In the pentagon, the edge lengths to the length of the diagonal are according to the golden ratio. And even the diagonal itself can be divided into two parts whose ratio corresponds to the golden ratio, if one chooses the point of intersection with other diagonals as a division point (◻ Fig. 8.13a). All diagonals together form the famous pentagram, whose meaning has undergone frequent changes in the course of cultural history. For example, it was a symbol for the planet Venus, for the goddess Venus, which was used by the Freemasons as a symbol for the five virtues of wisdom, justice, strength, moderation, and diligence, and from the Middle Ages, it was regarded as a sign against evil, while an upright pentacle is understood as a sign of occultism/satanism.

Although the icosahedron has triangular faces instead of pentagonal ones as boundary faces, the golden ratio can also be discovered here: the twelve corner points of the icosahedron form the corners of three equally large rectangles being oriented perpendicular to one another, which form golden rectangles, because their edge length ratio is also Φ (◻ Fig. 8.13b).

◻ **Fig. 8.12** Arrangement of leaves according to the golden section or angle. (Redrawn after Wolfgang Beyer on Wikipedia, ▶ https://de.wikipedia.org/wiki/Goldener_Schnitt#/media/File:Goldener_Schnitt_Blattstand.png)

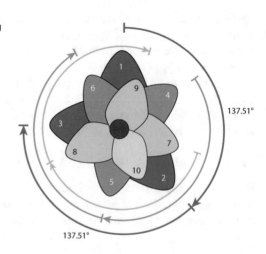

(a)   **(b)**

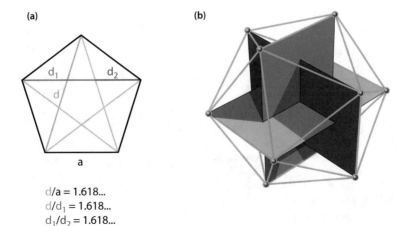

$d/a = 1.618...$
$d/d_1 = 1.618...$
$d_1/d_2 = 1.618...$

ロ Fig. 8.13 (**a**) The golden section in the regular pentagon: the ratio of the edge length to a diagonal, the ratio of the total diagonal to the long section, and the ratio of the long to the short section of the diagonal each give the golden section. (**b**) The three perpendicular inner rectangles of the icosahedron also have an edge length ratio corresponding to the golden section; they represent golden rectangles

Incommensurable Structures
In addition to quasicrystals, there are two other classes of aperiodic crystals that can be described using the concept of higher-dimensional crystallography or superspace crystallography:
1. The so-called composite crystals, in which each of two components forms their own lattice and these two lattices are incompatible with each other.
2. Incommensurably modulated structures.

Without going into too much detail, these two interesting classes should be briefly introduced to give you an idea of their structure.
1. *Composite Crystals*
   Imagine a substance made up of hexagonal channels consisting of hexagonal segment discs, which are primitively stacked along the crystallographic *c* axis. For example, rod-shaped guest molecules can now be incorporated into the center of these channels of the host structure, aligning with their longitudinal axis parallel to the *c* axis and forming "infinite" chains within the channels, always directly one behind the other and arranged regularly (ロ Fig. 8.14a). Now, if the distance from one guest molecule to the next has a different periodicity than the distance between the hexagonal segments that make up the channels (ロ Fig. 8.14b), then there are two possibilities: either the periodicities are such that there is a common multiple, then the two sublattices of the host and guest are not identical, but compatible with each other, e.g., by choosing a new superlattice, to which both partial lattices are in a rational relationship, or the two partial lattice constants *c* form an irrational relation, then this common multiple does not exist, and then the lattices are incompatible: they are incommensurable.
2. *Incommensurably Modulated Structures*
   Unlike in composite crystals, in incommensurably modulated structures, the disturbance of the regular crystal structure, which can be described by three lattice vectors, by a component with incompatible periodicities is an integral component of the structure. Let's look again at an example: Let's start with a disklike molecule that forms a regular lattice, here, for the sake of simplicity, reduced to two dimensions (ロ Fig. 8.15a). Now let's look at hypothetical polymorph, in which the centroid positions of the molecule are identical, but

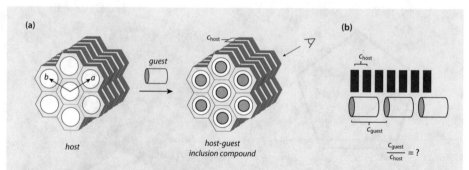

Fig. 8.14 Host-guest-inclusion compound (**a**), where the periodicity along the *c* axis is different for the host and guest (**b**) (see text)

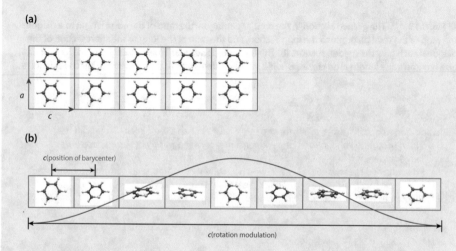

Fig. 8.15 (**a**) A molecule having a strict long-range order with regard to its position and orientation; (**b**) an example of a modulated structure in which the same molecule has a strict long-range order with regard to its position, but which rotates from cell to cell about its longitudinal axis

the molecule is slightly rotated from one lattice point to the next lattice point along the longitudinal axis (**▣** Fig. 8.15b). Then this rotation period, after which a full rotation around the longitudinal axis is given, may again form a periodicity that is incompatible with the periodicity of the centroid position along one of the lattice vectors.

For readers wishing to further explore this area of crystallography, the article in reference [17] is a good starting point.

Both phenomena of incommensurability, be it composite crystals or modulated structures, are accompanied in X-ray diffraction experiments with so-called satellite diffraction points, weaker exposure points around the main diffraction points. With the concept of crystallography of superspace, it is possible to indirectly derive the modulation frequency from the positions and intensities of these satellite reflections.

References

1. Reinhardt K (1918) Über die Zerlegung der Ebene in Polygone. Dissertation. Johann Wolfgang Goethe Universitat, Frankfurt a.M
2. Kershner RB (1968) On paving the plane. Am Math Mon 75:839–844. https://doi.org/10.2307/2314332
3. James RE (1975) New pentagonal tiling. Reported by Gardner, M. Sci Am 233:117–118
4. Stein R (1985) A new pentagon tiler. Reported by Schattschneider, D. Math Mag 58:308
5. Mann C, Mc-Loud-Mann J, Von Deroux D (2015) Convex pentagons that admit *i*-block transitive tilings. eprint arXiv:1510.01186. https://arxiv.org/abs/1510.01186. Accessed 30 Sept 2019
6. Shechtman D, Blech I, Gratias D, Cahn JW (1984) Metallic phase with long-range orientational order and no translational symmetry. Phys Rev Lett 53:1951–1954. https://doi.org/10.1103/PhysRevLett.53.1951
7. Mackay AL (1982) Crystallography and the Penrose pattern. Physica A 114:609–613. https://doi.org/10.1016/0378-4371(82)90359-4
8. Levine D, Steinhardt PJ (1984) Quasicrystals: a new class of ordered structures. Phys Rev Lett 53:2477–2480. https://doi.org/10.1103/PhysRevLett.53.2477
9. The Nobel Prize in Chemistry 2011. http://www.nobelprize.org/nobel_prizes/chemistry/laureates/2011/. Official web site of the Nobel prize. Accessed 30 Sept 2019
10. Steurer W (2018) Quasicrystals: what do we know? What do we want to know? What can we know? Acta Cryst A 74:1–11. https://doi.org/10.1107/S2053273317016540
11. Prof. Dan Shechtman 2011 Nobel Prize Chemistry Interview with ATS. https://www.youtube.com/watch?v=EZRTzOMHQ4s. Accessed 30 Sept 2019
12. Kuhn TS (1996) The structure of scientific revolutions.3rd edn. University of Chicago Press, Chicago
13. Bindi L, Steinhardt PJ, Yao N, Lu PJ (2009) Natural quasicrystals. Science 324:1306–1309. https://doi.org/10.1126/science.1170827
14. Bindi L, Yao N, Lin C, Hollister LS, Andronicos CL, Distler VV, Eddy MP, Kostin A, Kryachko V, MacPherson GJ, Steinhardt WM, Yudovskaya M, Steinhardt PJ (2015) Natural quasicrystals with decagonal symmetry. Sci Rep 5:Article number 9111. https://doi.org/10.1038/srep09111
15. Penrose R (1974) The role of aesthetics in pure and applied mathematical research. Bull Inst Math Appl 10:266–271
16. Takakura H, Gomez CP, Yamamoto A, de Boissieu M, Tsai AP (2007) Atomic structure of the binary icosahedral Yb-Cd quasicrystal. Nat Mater 6:58–63. https://doi.org/10.1038/nmat1799
17. Wagner T, Schönleber A (2009) A non-mathematical introduction to the superspace description of modulated structures. Acta Cryst B 65:249–268. https://doi.org/10.1107/S0108768109015614

Porous Crystals, Crystal Structures as Networks, and an Insight into Crystallographic Topology

© Springer Nature Switzerland AG 2020
F. Hoffmann, *Introduction to Crystallography*, https://doi.org/10.1007/978-3-030-35110-6_9

So far, in this book we have looked at a number of different classes of crystals that had one thing in common: they were invariably *dense* structures. This does not necessarily mean that their components completely fill the space, i.e., to 100%. For instance, it is well-known that the *densest* packing of spheres, whether these are atoms or cannonballs, fills the volume to only 74%. There are two of these densest packings: the cubic face-centered (cubic closest packing, ccp) and the hexagonal closest packing (hcp). A large class of crystals that can be described with these two sphere packings forms the structures of many metals, i.e., metallic *elemental crystals* (▶ Sect. 4.2.4). With a space-filling of approx. 68%, the cubic body-centered spherical packing is already somewhat less dense. This structure is also formed by some metals, for instance, V, Nb, Ta, Cr, Mo, W, and α-Fe.

We also already got to know two elemental crystals of nonmetals, graphite and diamond, in ▶ Sect. 7.4. Crystalline elemental crystals of nonmetals and compounds of nonmetals are much more complicated than crystals of metals because the *directed* chemical bonds that hold the atoms together in these structures allow for many more spatial variations of atomic arrangements than simply stacking spheres. Extensive *network structures* may be realized, e.g., the continuous cubic network of diamond or the continuous two-dimensional hexagonal network of the graphene layers in graphite. But these nonmetal (element) structures are densely packed, too; there are no major gaps or voids in these structures.

Many *ionic* (salt-like) crystals can be derived from sphere packings. These structures are often even denser than the structure of pure metals. Usually the smaller spherical cations occupy the tetrahedral or octahedral voids of the dense or densest packing of the (larger) anions. Also for this class of compounds, we have already encountered a number of examples; one of the best known is rock salt, in which the smaller sodium ions occupy the octahedral voids of the cubic closest packing formed by the (larger) chloride anions.

Another class of crystals is formed by *molecular crystals*, for which the structure of benzene was an example (▶ Sect. 7.2.). For these compounds the effective filling of space plays an important role, too; however, the specific structure, for instance, the mutual orientation of the molecules to each other, is determined by a multitude of different forces (van der Waals forces, π-π stacking interactions, classical and nonclassical hydrogen bonds).

With regard to the principle of space-filling, one could say that nature generally endeavors to avoid structures with pronounced voids. The reason for this is that the creation of surfaces or interfaces requires energy. Therefore, nature tends to realize structures that are as compact as possible by minimizing the surface area and thus reducing energy. However, this principle does not exclude the existence of porous structures with cavities or channels, as long as they are kinetically stabilized, i.e., the transformation into the more compact form with lower energy is inhibited and takes a very long time (sometimes several million years). Porous volcanic rocks, cork, and biological sponges are examples of natural representatives of porous structures.

Subsequently, we will discuss two prominent representatives of porous crystalline compounds, namely, zeolites and metal-organic frameworks. It is intended to demonstrate that it can be useful to look at these structures not only from a crystallographic point of view but to consider them at a slightly more abstract level as (chemical) *networks*. In the subsequent sections, we want to elucidate the characteristics of network structures, which different types exist and which descriptors are used to adequately describe their structures. First, the two mentioned compound classes will be briefly presented.

9.1 Zeolites and Metal-Organic Frameworks

9.1.1 Zeolites

The naturally occurring zeolites (gr. zeein = boiling, gr. lithos = stone) were discovered in 1756 by the Swedish mineralogist Axel Frederic von Cronstedt. Zeolites are formed as secondary minerals in volcanic areas permeated by hydrothermal water veins. Initially, they were regarded as mineralogical curiosities, and for the next 200 years, they were almost forgotten. It was not until the 1950s, when it was discovered that zeolites exist in quite considerable and degradable quantities. The first systematic investigations of their properties began, and also first attempts for their synthetical production started. Zeolites have gradually become one of the most important industrial chemicals and are now produced in a quantity of approximately one million tons per year. The most important technical application is catalysis: Zeolites are used for cracking of medium-heavy crude oil fractions and for the catalytic conversion of methanol into gasoline. But also beyond that, the spectrum of other fields of application is remarkable. For example, zeolites are used as ion-exchange resins for water softening, as a filler in paper and cement; they are used to separate gas mixtures, e.g., oxygen from air, furthermore for the purification of radioactively contaminated waste water, in flue gas desulfurization, as an abrasive in toothpaste, for the regeneration of dialysis fluids and motor oil, as adsorbents in cigarette filters, as a drying agent, and also as a catalyst in organic synthesis chemistry.

With 48 naturally occurring representatives, zeolites form one of the largest mineral groups. In the meantime, more than 150 further zeolites have been synthesized, at least on a laboratory scale. Zeolites belong to the so-called tectosilicates ("framework silicates"), which are composed of anionic (negatively charged) SiO_4 and/or AlO_4 tetrahedra and positively charged counterions (often sodium, potassium, or calcium). What distinguish the different zeolites are their different cavities: pores and canal systems which pass through the structure. These are often filled with water in their original state, but it can be removed by drying procedures (e.g., by applying vacuum or heating up the material). The pores or cavities of zeolites are of different sizes, have different shapes, and are connected to each other in different ways. In ◻ Fig. 9.1 two different zeolite types are shown, on the one hand the Linde type A zeolite (LTA) and on the other hand the sodalite type (SOD).

How can these structures be best described or represented? One possibility would be exactly the type of representation chosen in ◻ Fig. 9.1a, b, which is only the visual representation of the atomic coordinates and the chemical bonds between them. In a further analysis, we could now identify the building units and would discover that both structures consist exclusively of Al/SiO_4 tetrahedra and in both cases all linked tetrahedra have exactly one common vertex, while common edges or faces do not occur (◻ Fig. 9.1c, d). This is the case not only for these two zeolites but also for all of the 231 known zeolite types, which are deposited in a publicly accessible database. This database is maintained by the International Zeolite Association (IZA). It regularly publishes the *Atlas of Zeolite Framework Types* [1]. However, it is now clear that a representation by coordination polyhedra cannot be a differentiating description for this class of substances. As already mentioned above, the cavities and channels are the characteristic features of zeolites. The boundaries of the pores are formed by cyclic atomic chains,

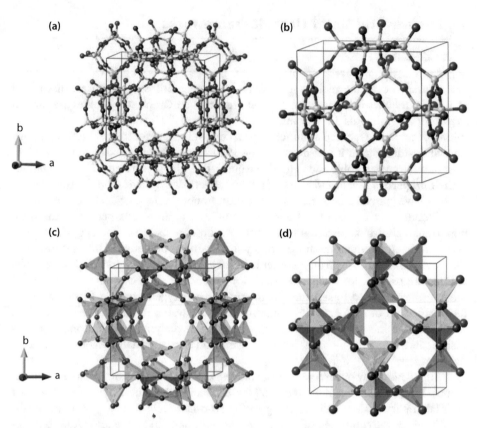

■ **Fig. 9.1**   Atomistic ball-and-stick model of the zeolite Linde type A (LTA) (**a**) and sodalite (SOD) (**b**) as well as a representation as coordination polyhedra of the two models in which the Al/SiO$_4$ tetrahedra are highlighted (**c**), (**d**). For the sake of simplicity, the cations required for charge neutralization have been omitted

rings of a certain size and arrangement. And that puts us already in the middle of the domain of network descriptions. In the case of zeolites, we first disregard the concrete atomic structure by omitting the oxygen atoms, which only link two tetrahedra but are not *branching points*. What remains is a network of tetrahedrally coordinated aluminum or silicon atoms, which are therefore also called T atoms. They represent the *nodes* of the network that are connected by *edges* that are rendered as lines. The result of this abstraction process, also called *deconstruction*, for the two zeolites LTA and SOD is shown in ■ Fig. 9.2. This type of representation makes it easier to identify the larger cage-like units, which in turn consist of *secondary building units* (SBUs) (the primary building blocks are simply the Al/SiO$_4$ tetrahedra). The zeolite LTA is composed of double-four rings (in the form of a cube), β or sodalite cages (the largest cage in sodalite, which has the shape of a truncated octahedron), and the even larger α or supercages (truncated cuboctahedron) (■ Fig. 9.3). Sodalite consists exclusively of sodalite cages arranged in a body-centered cubic packing. In addition to the SBUs and cages, there are further topological descriptors to fully characterize these networks. These are described in detail in ► Sect. 9.3.

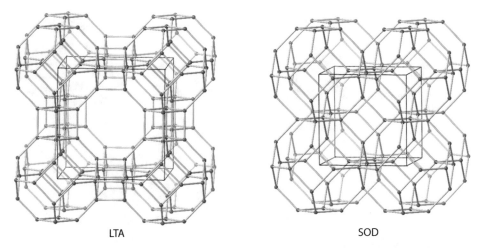

◘ Fig. 9.2   Representation of the zeolites Linde type A and sodalite as a net, in which the T atoms (in blue) form the nodes, which are connected to each other by edges (in yellow). The unit cell is drawn in black

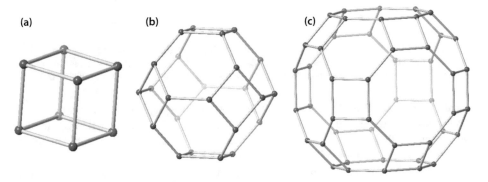

◘ Fig. 9.3   The building blocks of the zeolites LTA and sodalite: **(a)** double-four ring (left); **(b)** β or sodalite cage; and **(c)** α or super cage

9.1.2 Metal-Organic Frameworks

Metal-organic frameworks (MOFs) form a relatively new class of porous crystals that are purely synthetic in nature; at least they have not yet been discovered in nature. MOFs consist of two components: an inorganic building block consisting of metal atoms or metal-oxygen clusters (*connector*) and an organic building block (*linker*) bridging the inorganic building blocks. They form infinitely extended coordination networks, which generally represent a relatively rigid framework. In ◘ Fig. 9.4 you can see one of the most famous MOFs, the prototypical MOF-5, and its assembly of the individual components, which consist of a Zn_4O tetrahedron and the dianion of terephthalic acid as a linker [2]. The terephthalic acid dianion coordinates with both oxygen atoms, thus bridging two zinc atoms each. Overall, six linkers coordinate on a Zn_4O tetrahedron (a tetrahedron has six edges) in octahedral fashion. By connecting the carbon atoms of the six linkers that coor-

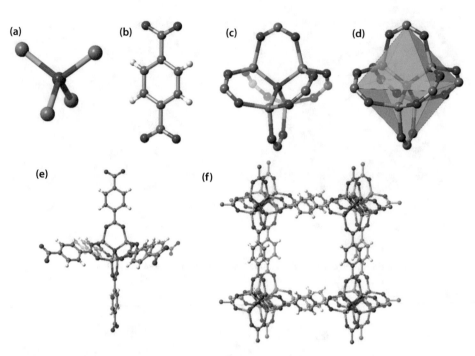

Fig. 9.4 Structure of MOF-5; (a) inorganic tetrahedral Zn_4O unit; (b) terephthalate dianion, organic bridging ligand (linker); (c) coordination sphere around Zn_4O unit; (d) green octahedron determining the shape of connector; (e) six linkers bind to a Zn_4O unit; (f) section of the primitive cubic network of MOF-5

dinate to the Zn_4O unit, they define the coordination polyhedron of the connector (■ Fig. 9.4d). These transition points from the inorganic SBU to the actual bridging moiety are also called points of extension. Overall, an infinitely extended, cubic-primitive framework forms. Compared to zeolites, MOFs are characterized by the fact that the relative void volume can assume much larger values (70% are not uncommon) and that their structural diversity significantly exceeds that of the zeolites. The latter is due to the fact that the inorganic structural unit can have very different coordination numbers as well as geometries, whereas in the case of zeolites exclusively tetrahedrons occur; and also the organic bridging ligand contains additional branching points, while the bridging oxygen atom of the zeolites always connects only two Al/SiO_4 tetrahedra. A selection of representative geometries of connectors and structures of linkers is shown in ■ Figs. 9.5 and 9.6, respectively.

Although the class of MOFs has been developed only since the late 1990s, meanwhile more than 35,000 different structures are known. And also the number of network types behind these structures has grown enormously, especially in the last decade, to approx. 3,000.

The large void volume and the very large internal specific surface area – 1 gram of these materials can have an internal surface of up to 7,000 square meters, comparable to the area of a soccer field – make MOFs suitable for a number of applications, particularly for gas storage. The chemical company BASF is currently producing MOFs on a tonne scale in order to evaluate them in a field trial for their suitability as storage systems for natural gas tanks in trucks. There is a video on YouTube worth watching, in which one of the active researchers in the field of MOFs explains the principle of the operation [3].

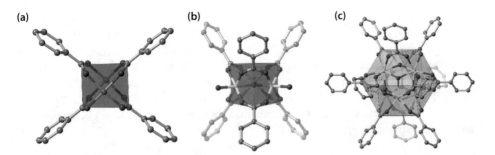

◼ **Fig. 9.5** Selected connectors frequently found in MOFs: (**a**) square-planar geometry with four bound linker molecules in the form of a so-called paddle-wheel motif, (**b**) trigonal-prismatic geometry with six bound linkers, and (**c**) cuboctahedral coordination of 12 linker molecules

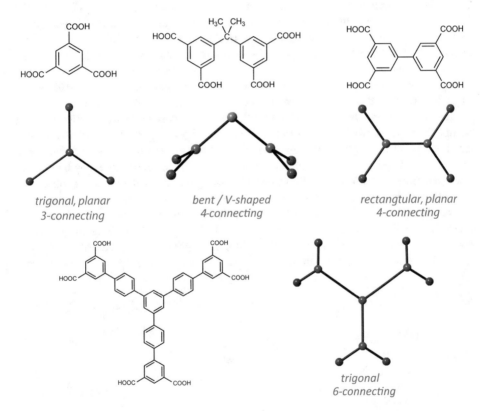

◼ **Fig. 9.6** Selection of some representatives of bridging ligands used for the synthesis of MOFs and their abstract representation as nodes connected by edges (drawn as black lines) as well as their point of connections

The process of abstraction, which turns the concrete atomic structure of a MOF into its representation as a network, is somewhat more complicated compared to zeolites, because very different coordination numbers and forms are involved and because it is not always clear how the organic component is best represented as a node. But the principle is always to analyze how many linkers originate from an inorganic node in which geometry and how many branching points of which geometry the linker has (see also ◘ Figs. 9.5 and 9.6). Once a net has been made from zeolites and MOFs, i.e., a thing consisting of vertices and edges, the next step is to describe this net. How many different vertices and edges are there? What do we mean by "different" in this context? What is the coordination number of the vertices, and which types of vertices are connected via edges to which other vertices? Is it sufficient to specify the coordination number for vertices? We will see that this is not the case. Additionally, we need information about the spatial linkage of the vertices – and that's where *topology* comes into play.

9.2 Introductory Remarks on Networks and Graphs and Their Topology and Symmetry

Networks, or simply nets in short, are not only used in crystallography or by chemists. We encounter them everywhere in everyday life. Be it a plan from the subway network, electronic circuits, the indispensable connection to the Internet via the Wireless Local Area Network (WLAN), our brain as a neural network or artificial replicas of it, or the ever-increasing involvement in social networks, like Facebook, YouTube, Twitter, and Instagram. All these structures have in common that some parts of them can be identified as nodes and that these nodes have connections to other nodes. These interconnected nodes can be represented as special *graphs* that are purely mathematical entities: abstract points, the vertices, which are connected by lines or edges. In general, we can distinguish different classes of graphs, which are characterized by certain properties. For example, a graph may contain loops (vertices that are connected to itself), or there may be vertices that have more than one connection to a neighboring vertex, or there may be loose ends (lines that do not lead to any other vertex at the other end). Furthermore, there are networks where the direction between two vertices is decisive – for example, in a flowchart – and there are finite and infinite graphs or nets. In ◘ Fig. 9.7 three different graphs are shown. However, if we are dealing with zeolites or MOFs, we are faced exclusively with graphs that have no loops and no loose ends (a crystal is, in principle, an infinite struc-

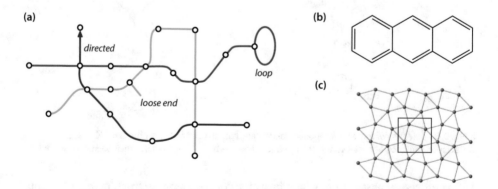

◘ **Fig. 9.7**   Some examples of different types of graphs. (a) A finite, non-simple graph with directed edges, a loose end and a loop; (b) a finite, non-simple graph with multiple bonds between some of the nodes; (c) a simple, infinitely extended net with translational symmetry in the plane (the black frame marks the unit cell). (Reprint courtesy of Wiley-VCH, from [3])

ture). Furthermore, there are no multiple connections between vertices, even if the chemical bond between components of a network should have a multiple bond – this is not taken into account, important is only whether the building blocks are interconnected or not. In addition, there are no directional edges. Graphs that satisfy all these conditions are called *simple graphs*. The networks of MOFs can have translational symmetry in two or three spatial directions, those of zeolites are exclusively those with translational symmetry in three spatial directions.

The Seven Bridges of Königsberg

Leonhard Euler (one of the most important mathematicians of all times, * 15.04.1707 Basel, † 18.09.1783 St. Petersburg) solved the famous problem of the seven bridges of Königsberg with a graph-like approach. He is, therefore, regarded as the founder of the mathematical discipline graph theory, which in turn is a branch of topology. Here is the story: The river Pregel flows through the city of Königsberg (today's Kaliningrad), partly also two-armed, dividing the city in four parts (until about 1945). A total of seven bridges connected the city districts. The question to be answered was: Is it possible to reach all four parts of the city in succession from a certain point and arrive at the same starting point without having to use a bridge several times? In other words, is there a way through the city where you walk over *all* the bridges, but only *once*? In ▢ Fig. 9.8a a schematic sketch of the course of the river, the bridges, and the districts is shown and in ▢ Fig. 9.8b the completely abstract graph with points (districts) connected by edges or lines (bridges). The answer to the question is: No, such a route does not exist. The idea that Euler used to solve the problem was as follows: *All* parts of the city are connected by an *odd* number of bridges. For example, if there is only one bridge to the middle island (No. ② in ▢ Fig. 9.8a), it is clear that we need to take the same bridge to get away from the island, so the undertaking already fails. If there are two bridges connecting two parts of the city, then we could use one for entering the district and the other for leaving. But if there are three… We can further deduce: if we have four quarters, then a maximum of two quarters may have connections to other parts of the city with an odd number of bridges. The rest would have to have an even number of bridges to leave on a new path. The bridge problem is not a classic *geometrical* problem because it does not depend on the precise location of the bridges but only on the connection scheme of bridges connecting the districts. It is therefore a *topological* problem.

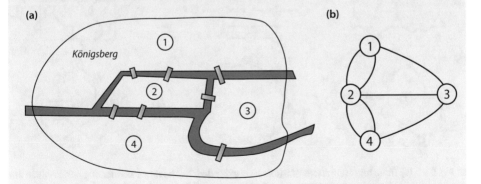

(a)   **(b)**

▢ **Fig. 9.8**   (a) Schematic representation of Königsberg (today Kaliningrad), which was separated into four districts by the Pregel River until 1945. A total of seven bridges crossed the river. (b) Graph-theoretical representation of the districts ① to ④ (nodes), which are connected by a different number of bridges (edges)

9.2.1 Topology

In general, topology (gr. *topos* = place and gr. *logos* = study) is a fundamental subdiscipline of mathematics which is concerned with properties of mathematical structures in space or objects in space (i.e., set of points) that do not change under continuous deformation. This continuous deformation includes bending and stretching or distortion, but it excludes operations such as sticking previously unconnected parts of the space or objects together. The opposite is forbidden, too, breaking apart of formerly connected parts. For example, if we look at the net in ◻ Fig. 9.8b, topologically it does not matter whether the four nodes (representing the neighborhoods) are 1,000 km apart, or just 10 mm. In fact, mathematically viewed, the edges do not have a property called length. It is also immaterial whether the four nodes form a symmetrical shape or are arranged completely oblique to each another. As long as two arrangements of nodes can be transformed into each other by continuous deformation (without breaking up connecting lines), they are topologically identical. In ◻ Fig. 9.9a you can see an example of three nets, which are geometrically different but topologically identical, because they are interconvertible according to the rules of topology. In contrast, it is not sufficient to characterize a network by indicating which nodes are connected to how many and to which others. This is visualized in ◻ Fig. 9.9b by three graphs or nets, which have identical linking patterns[1] – they are *isomorphic* – but different topologies, because they are not interconvertible without breaking at least one edge and then reassembling it again. This identity or nonidentity of topological entities leads to two further questions that are closely related:

— How should a network be represented?
— On which basis can the question be answered as to whether two networks are identical?

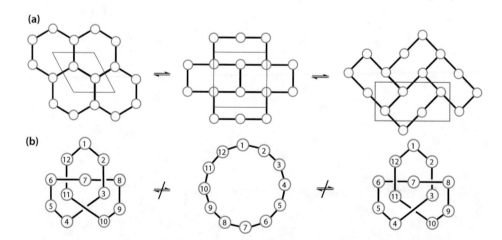

◻ **Fig. 9.9**   **(a)** Three representations of a (infinitely extended) net having 3-connecting nodes which are geometrically different but topologically identical (the unit cell is drawn in red). **(b)** Three identical (isomorphic) graphs, in which all nodes always have an identical connection but different topological realizations. (Reprinted with kind permission by Verlag Wiley-VCH, from [3])

---

1   In all three networks, all nodes have two neighbors; node 1 is always connected to nodes 12 and 2, node 2 to 1 and 3, and so on.

Only if the latter question is answered, we can begin with a reasonable classification or identification of prototypes of nets. To give an example, in a topological sense, vertices have no concrete location in space; they are abstract, infinitely small objects that can be placed anywhere in space, provided that their connection scheme remains identical to all other vertices. Therefore, there is an agreement of how to assign vertices to coordinates in space: it should be done in such a way that a minimum number of vertices and edges of different symmetry arise. Crystallographically speaking, the vertices and midpoint of the edges should be placed at coordinates with a maximum site symmetry (▶ Sect. 6.1). Vertices and edges that are symmetry-related to each other are identical. The procedure to assign coordinates to the vertices (and hence edges) of an abstract graph is referred to as an *embedding* or also a *realization*. If we now refer to ◘ Fig. 9.9a again, we see that in all three nets, all nodes have the coordination number three (this is abbreviated as "3-c"). The ring size is identical, too: only six-membered rings are involved, although we still have to define exactly what is actually meant by a ring. But if we look at the symmetry, we immediately see that the leftmost hexagonal net in ◘ Fig. 9.9a is the embedding with the highest symmetry, while the other two nets have lower symmetry realizations. This can be easily inferred from the number of nodes per unit cell, which gives the lowest value for the hexagonal network, namely, two, while the other two networks each have four nodes per unit cell.

The process of maximizing symmetry or spatial arrangement of nodes connected by edges and the subsequent examination which type of net is present can only be done manually for relatively simple structures. For complicated structures, modern software packages are being used, the most important ones are *Systre* [4] (a subprogram of the more comprehensive GAVROG package) and *ToposPro* [5].

Before we take a closer look how networks are characterized, we will consider how nodes of identical coordination number can actually result in topologically different structures.

9.2.2 Interplay of Local Symmetry and Global Topology

First of all, nodes of identical coordination number, regardless of their coordination geometry (see ◘ Fig. 9.9a), are topologically identical. For example, tetrahedral and square-planar coordinated nodes are identical nodes – topologically, they are simply 4-c nodes. But this raises the question how networks of different topologies can result on the basis of topologically identical vertices. If we only consider the nets of MOFs, then there are already more than 220(!) different nets consisting exclusively of nodes that are 4-coordinating. Although the local geometry of the single vertices involved in the net is not a topological feature per se, it is exactly this geometry which gives rise to the different possibilities to arrange a set of interconnected vertices in space, which are topologically distinguishable (cannot be transformed into each other without breaking bonds). It is the interplay of the local geometry of the topological neighborhood of the vertices together with their relative orientation to each other that leads to globally different topological arrangements of the vertices (see ◘ Fig. 9.9b). It may not be a problem to see that two different networks result if, in one case, only one kind of tetrahedrally coordinated node is used and, in the other case, only square-planar nodes are involved (◘ Fig. 9.10a, b). However, it is a little less obvious that it is possible to realize different networks even if only one single type of node is used. The two nets in ◘ Fig. 9.10b, c each consist of only one kind of planar 4-c nodes – such nets that consist of only one kind of node are called

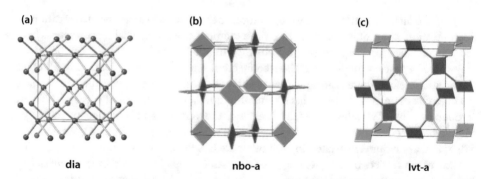

(a)   (b)   (c)

dia   nbo-a   lvt-a

◻ Fig. 9.10   Three networks made up exclusively of 4-connecting nodes: (a) the **dia** network with tetrahedrally coordinated nodes; (b) the **nbo**, and (c) the **lvt** networks, both constructed of square-planar nodes. To clarify the coordination geometry, the nodes of the **nbo** and the **lvt** net are drawn as filled squares; this kind of representation, in which the nodes are replaced by their respective coordination polygons or polyhedra, is called augmentation (lat. augmen = proliferation, multiplying) and is identified by the addition -**a**. For an explanation of the symbols, see ▶ Sect. 9.3. (Reprinted with kind permission of Verlag Wiley-VCH [3])

uninodal – and yet they are topologically distinct. Up to this point, we cannot say in which way they are different. To do so, we obviously need other descriptors. We will take a closer look at them in the following section, and we will also learn how to uniquely identify and name networks.

9.3 Descriptors and Nomenclature of Nets

The description or characterization of nets can be a rather complicated matter. Two-dimensional periodic nets consisting of only one type of nodes are still relatively simple. Imagine that you are a fisherman and you would have to describe the Fishing Regulatory Authority which nets you use to catch fish. You state the following: if I stretch the net to all sides, the meshes will look square and have an edge length of 15 mm, and from all nodes four threads go out to four neighboring nodes. This is not too complicated yet, but you can easily imagine the considerable effort that is needed, if we switch to three-dimensional periodic nets with different types of nodes. Therefore, it makes sense to consider a uniform system that specifies which properties of a net should be determined, i.e., which features are its characteristics. In addition, it makes sense to devise a suitable naming system which, in practical terms, refers to a net by means of a unique name instead of having to specify all of its characteristics. The fishing net, for example, could have the name 15Q4. This section will explain which characteristics are used to describe chemical nets and which nomenclature systems exist for them.

9.3.1 The RCSR Symbol and the RCSR Descriptors

To clearly identify a network in the field of coordination polymers and MOFs, a three-letter code was invented to act as an identifier. These three letters are set small and bold; **dia** (derived from diamond) would be an example. All the different networks are collected in a database called the Reticular Chemistry Structure Resource (RCSR) [6], which is freely accessible online and in which the essential characteristics of a network are retrievable. This is a remarkable project initiated by Michael O'Keeffe from the Arizona State University, who has built up and maintains this database. The database meanwhile includes about 2,900 three-dimensional and 200 two-dimensional periodic networks as well as 120 polyhedra, which can also be represented as a network.

The procedure of assigning each net a distinctive three-letter code is analogous to the practice of the long-established database for zeolites (▶ Sect. 9.1.1). The only difference is that the abbreviations for the zeolite types are uppercase (e.g., SOD for sodalite, FAU for faujasite, or MOR for mordenite). The names for some nets in the RCSR are derived from structure types or minerals (e.g., **dia** = *diamond*, **pcu** = *primitive cubic*, **sql** = *square lattice*, **qtz** = *quartz*, **ant** = *anatase*, **tbo** = *twisted boracite*). Others are a kind of short symbol of the formula of the corresponding compound, in which either one or more atoms form the corresponding network (e.g., **cds** = the network of cadmium *and* sulfur atoms in $CdSO_4$, **crb** = the network of boron atoms in CrB_4, **srs** = the network of silicon atoms in $SrSi_2$, **pts** = the network of platinum *and* sulfur atoms in PtS). However, most three-letter symbols are randomly assigned and do not refer to any structure or name.

Although these three-letter codes clearly identify a network, we do not know yet which specific features this or that network has. In order to do so, we can retrieve the respective network in the RCSR database to inspect the net specifications. The most important entries that we find under an entry will be explained below using the example of the **dia** network. It is advisable that you visit the corresponding page on the Web to understand the explanations. You can do this by going to the start page of the database for three-periodic networks (▶ http://rcsr.net/nets), typing "dia" in the field "Symbol" and clicking the "Search" button; alternatively you can directly access the desired page at the URL ▶ http://rcsr.net/nets/dia. A screenshot of the entry for the **dia** net is shown in ◻ Fig. 9.11.

9.3.1.1 Picture, Names, Keywords, and References

For many nets you will find a pictorial representation of the net directly below the symbol (here **dia**). This can correspond to different styles, for example, a ball-stick or tiling model. Directly below is also the direct link to the entry of the **dia** network. If the three-letter code is derived from a name or a structure, this is listed here (in this example "diamond"), along with alternative names or symbols that other sources use for this net (here "sqc6" and "4/6/c1"; the latter is the Fischer symbol, which will not be explained here). Then a line with associated keywords and a reference may follow. However, not every network has such a reference.

9.3.1.2 Embeddings: How to Assign Coordinates to Vertices?

The next three tables specify properties of the network that refer to the following two points:

- The relationship between an abstract-mathematical entity, the graph of the net, and a real object that has a certain spacious dimension and consists of building blocks that have certain distances to each other, etc.
- Furthermore, the question arises which guidelines should be followed when assigning vertices (which either represent an SBU or a branching point of a linker) to coordinates in 3D space, i.e., the question already mentioned above regarding the embedding of the graph in three-dimensional space.

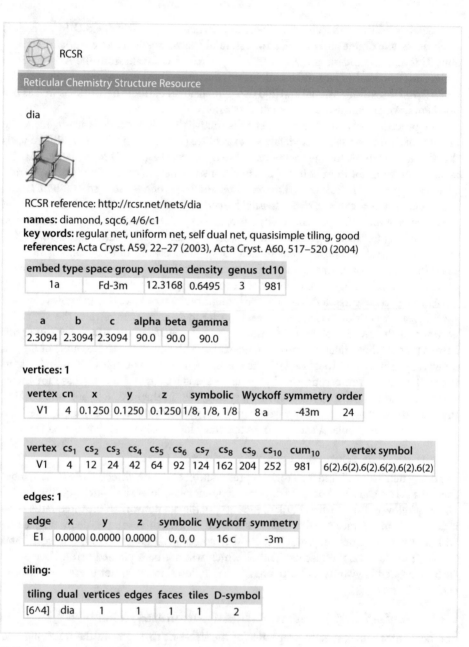

RCSR

Reticular Chemistry Structure Resource

dia

RCSR reference: http://rcsr.net/nets/dia
names: diamond, sqc6, 4/6/c1
key words: regular net, uniform net, self dual net, quasisimple tiling, good
references: Acta Cryst. A59, 22–27 (2003), Acta Cryst. A60, 517–520 (2004)

embed type	space group	volume	density	genus	td10
1a	Fd-3m	12.3168	0.6495	3	981

a	b	c	alpha	beta	gamma
2.3094	2.3094	2.3094	90.0	90.0	90.0

vertices: 1

vertex	cn	x	y	z	symbolic	Wyckoff	symmetry	order
V1	4	0.1250	0.1250	0.1250	1/8, 1/8, 1/8	8 a	-43m	24

vertex	cs_1	cs_2	cs_3	cs_4	cs_5	cs_6	cs_7	cs_8	cs_9	cs_{10}	cum_{10}	vertex symbol
V1	4	12	24	42	64	92	124	162	204	252	981	6(2).6(2).6(2).6(2).6(2).6(2)

edges: 1

edge	x	y	z	symbolic	Wyckoff	symmetry
E1	0.0000	0.0000	0.0000	0, 0, 0	16 c	-3m

tiling:

tiling	dual	vertices	edges	faces	tiles	D-symbol
[6^4]	dia	1	1	1	1	2

◘ Fig. 9.11 Screenshot of the entry of the net **dia** in the RCSR

For example, consider again the fact that an edge connecting two vertices can have any length, or strictly speaking "length" is not a (physical) property of the edge at all. Furthermore, a graph initially has no symmetry, at least no classical crystallographic symmetry. Therefore, concerning the procedure of embedding, i.e., assigning vertices real fractional coordinates, the following guidelines have been agreed upon:

1. All edges should be equal in length and have the value of 1 (in units of the lattice parameters).
2. The shortest distances between vertices should correspond to edges (because these represent bonds between the chemical building units).
3. The embedding should result in a structure with maximum symmetry.
4. The volume of the unit cell should be maximized, subject to the constraint of equal edge lengths, which means – vice versa – that the density (vertices per volume unit) should be minimized.

There are many networks where simultaneous compliance with all directives is impossible and compromises have to be made. The type of compromise is encoded as an embed type. Details of the encoding can be found on the "About" page of the database (▶ http://rcsr.net/about).

Following as many of these guidelines as possible, the crystallographic data are specified: the *space group* (note that always the origin choice 2 of the International Tables for Crystallography is chosen here), the *metric*, and the *volume* of the unit cell (a, b, c, α, β, γ, V). The *density* is given by the number of nodes per unit cell divided by its volume; in our **dia** example, it is $8/12.3168 = 0.6495$.

9.3.1.3 Genus

It may seem a bit strange, but even the surfaces of a topological object can be assigned a *genus*, whereby here not the usual dichotomous gender of humans or animals is meant. It simply forms something like a category here. In a strictly mathematical sense, genus is a rather complicated property of the surface of a topological object. In a less strictly mathematical sense, the genus of an object can be vividly described as the number of its "holes" or "handles." To give a few examples, a glass sphere, a football, or also a wine glass have the genus 0; a coffee cup, a swimming ring, or a donut have the genus 1; a pair of scissors and a double torus have the genus 2; and a pretzel (at least most of them) have the genus 3. Alternatively, the genus can also be defined as the maximum number of continuous cuts through an object, provided that the surface of the object is still a continuous piece after the cut. A cut through a ball would cut it into two separate pieces, so the genus is 0. The situation is different with a bicycle tube: a radial cut along the entire circumference would not cut it into two pieces; it would still consist of a continuous piece of rubber. But a second such cut would divide it into two pieces, so the genus is 1.

But now the question arises, how we can apply this mathematical concept to nets, because we are dealing with infinitely small vertices and edges, which per se have no surface at all. First, we have to inflate the vertices and edges to finite dimensions, at least virtually. Secondly, we have to consider that we are dealing with two- or three-dimensional periodic nets that have an infinite number of holes. Therefore, the genus is related to the smallest repeating unit of the net, i.e., to its primitive (not centered variants) unit cell. To determine the genus of the network, we proceed as follows: we take the nodes of the repeating unit together with their first neighbors. Then, we link the neighboring nodes, which start from the repeating unit in a positive crystallographic direction $+[uvw]$, with the neighboring nodes, which lie in the inverse crystallographic direction $-[uvw]$, by a handle (◻ Fig. 9.12). The number of handles thus created corresponds to the genus. If we remember that the vertices and edges were assigned finite dimensions in this procedure, we see that the number of handles corresponds to the number of holes in a continuous surface.

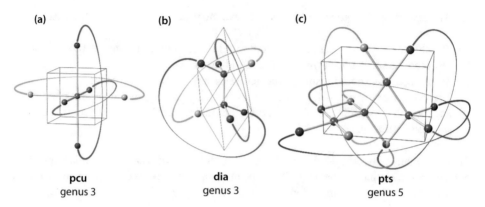

Fig. 9.12 The repeat unit (blue nodes) and its pairs of topological neighbors in the direction +[*uvw*] and -[*uvw*] of the networks **pcu (a)**, **dia (b)**, and **pts (c)**. The topological pairs are connected by a handle, forming a body with a corresponding number of handles; the number of handles corresponds to the genus of the net. (Reproduced with the kind permission of Verlag Wiley-VCH, from [3])

9.3.1.4 Topological Density (td10)

The topological density – represented by the *td10* value – is the accumulated number of neighboring vertices up to the tenth coordination sphere or shell (see also below, cum_{10} value) of a vertex, including the one reference vertex (coordination sphere 0). For the **dia** net, the td10 value is 981. For nets with more than one type of node, a weighted average is given. Keep in mind that a topological density has nothing to do with a physical density ("kilogram per cubic meter"), because firstly, nodes weigh nothing and, secondly, they have no physical distance from each other. Nevertheless, topological density is a property of networks that is used for their closer characterization.

embed type	space group	volume	density	genus	td10	deg freedom
1a	Fd-3m	12.3168	0.6495	3	981	1

a	b	c	alpha	beta	gamma
2.3094	2.3094	2.3094	90	90	90

9.3.1.5 Specification of the Vertices

The next entries within a datasheet for a network specify the vertices of the network (in its most symmetric embedding) with respect to the coordination number (cn), their fractional coordinates (*x, y, z*) in numerical and symbolic form, the Wyckoff symbols, multiplicities, and site symmetry. Due to the fact that we are dealing with periodic networks, the same descriptions are made here as in real, physical crystals; see ▶ Chap. 6. In addition, the order of the corresponding crystallographic point group is given, which is the number of all symmetry operations.

vertex	cn	x	y	z	symbolic	Wyckoff	symmetry	order
V1	4	0.1250	0.1250	0.1250	1/8, 1/8, 1/8	8a	-43m	24

In the second table of the vertex specification, two further entries are listed: the *coordination sequence* (cs) and the *vertex symbol*.

9.3.1.6 Coordination Sequence

The coordination sequence specifies the number of topological neighbors of a vertex from the first to the tenth coordination shell, that is, the number of vertices that can be found exactly one, two, three, etc. edges away from the parent node. Furthermore, the sum over all coordination spheres is recorded under the entry cum_{10}, which means $cum_{10} = \sum_{i=1}^{10} cs_i + 1$.

If there is only one kind of vertex in the network, as in this example, then the cum_{10} value is equal to the td10 value. The coordination sequence for the first four coordination spheres (cs4) of the network **sql** is illustrated in ◘ Fig. 9.13.

9.3.1.7 Vertex Symbol

The vertex symbol is one of the most important characteristics of nets. It is not absolutely specific for a net, i.e., there are different nets that can have the same vertex symbol, but this is the absolute exception. It is therefore important to understand how vertex symbols are determined, and how they are constructed. You also need this knowledge to be able to decide, for example, whether you have a new net in front of you, which you may have synthesized.

Vertex symbols can be derived for nets that are periodic in two or three dimensions, but the notation system of the vertex symbols is slightly different for these two cases. In

◘ **Fig. 9.13** Coordination sequence of the **sql** network up to the fourth coordination shell cs4: 4, 8, 12, and 16. (Reprinted with kind permission by Verlag Wiley-VCH, from [3])

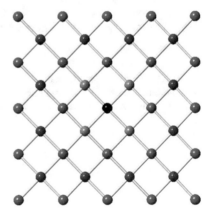

addition, there are slight differences for three-dimensional periodic nets, depending on the coordination number of the vertices. The general recipe for deriving the vertex symbol is as follows:

1. Inspect consecutively all angles of a vertex (each two edges which come together at a vertex). In two-dimensional periodic nets, there are exactly n angles for a vertex with a coordination number of n, while the number of angles in vertices in three-dimensional periodic nets is $n \cdot (n - 1)/2$. For example, a tetrahedral vertex with the coordination number four has six angles.
2. Examine in which shortest *rings* (the difference between any *cycle* and a *ring* is explained below, for the time being the author assumes that your intuitive understanding of the term is perfectly true) these angles are included, and note the size of the rings (number of vertices of the rings).
3. Separate the individual specified ring sizes with a point. This means that the general form of a node symbol is *a.b.c.d...* If rings of identical size occur one after the other, they may be combined in two-dimensional periodic nets in the form of a superscript index number, i.e., in the form $a^m.b^n.c^o.d^p$.
4. This procedure should be repeated for each vertex type (of different symmetry or coordination number).

The only question that remains to be answered is the order in which you should examine the angles and thus specify the rings. For two-dimensional periodic nets – here each angle is involved in only one ring (!) – there is a rule to proceed in *cyclic order*. We start with the angle involved in the smallest ring or in one of the smallest rings and then turn continuously either clockwise or counterclockwise. The direction is determined by the direction in which a sequence of numbers or ring sizes as small as possible results. The following example should provide clarity: The net **hnc** shown in ◨ Fig. 9.14 – composed of five-membered and seven-membered rings – consists of a total of four different vertices. Three of them get the vertex symbol 5.7.7, while one gets the symbol 5.5.7 (and not 5.7.5).

If we turn to three-dimensional periodic nets, the determination of the vertex symbols becomes more complicated. There are three reasons for this:

◨ **Fig. 9.14** The two-dimensional periodic, quadrinodal (3,3,3,3)-c net **hnc**. The unit cell is marked with a black frame

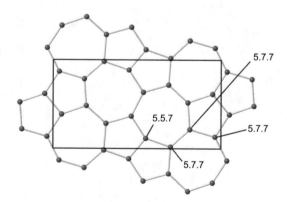

5.7.7

5.5.7

5.7.7

5.7.7

(a)   (b)

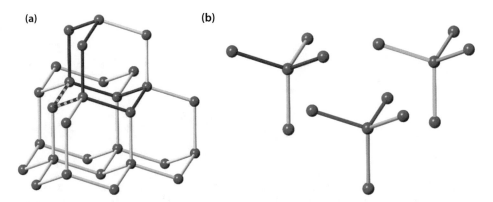

◩ Fig. 9.15 In three-dimensional periodic nets, an angle can be involved in more than one ring. (**a**) The red-grey-striped angle of the **dia** net, shown on the left, is involved in two rings simultaneously which are highlighted in red. (**b**) The six angles of the 4-connecting nodes are grouped into pairs of opposite angles, i.e., those that do not have a common edge. (Reproduced with kind permission of Wiley-VCH, from [3])

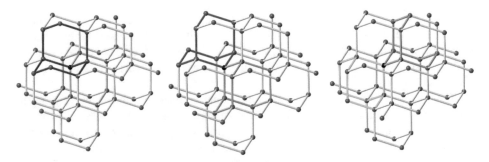

◩ Fig. 9.16 The three pairs of opposite angles of the **dia** net together with the rings in which they are involved. Each angle is involved in two rings simultaneously. (Reproduced with kind permission of Wiley-VCH, from [3])

First, one angle can be involved in several rings at the same time. For example, all of the vertices of the **dia** net are included in six-membered rings, but for each angle there are two six rings (◩ Fig. 9.15a). Rings of the same size at the same angle are now combined with a subscript index, i.e., for the gray and red dashed angle shown in ◩ Fig. 9.15a, the vertex symbol is 6_2. Unlike for two-dimensional periodic nets, similar angles are never grouped with another superscript index. Instead, all angles or rings of an angle are separated by a point.

Second, for vertices with a coordination number larger than three, it is not possible any more to determine a cyclic order of the angles. This raises the question in which order the angles should be examined. For 4-coordinating vertices, there is the agreement to group the six angles into three pairs of opposite angles, that is, those angles that have no common edge (◩ Fig. 9.15b). The three pairs of angles and the corresponding rings in which they are involved are highlighted in ◩ Fig. 9.16. There is only one kind of vertex, and the vertex symbol is $6_2.6_2.6_2.6_2.6_2.6_2$. For the sake of simplicity of presentation in com-

(a)   (b)   (c)

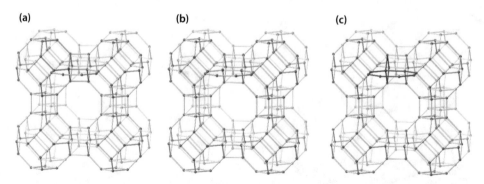

◻ **Fig. 9.17** All tetrahedral 4-c vertices of the **Ita** net are identical. Their vertex symbol is based on the ring size of the angles that come together at a vertex. Pairs of angles without a common edge are build. For each of these pairs, the smaller ring size is always specified first. For the three pairs of angles of the vertex highlighted in red, there are four- and six-membered rings for the first pair ((**a**), marked green), another four- and six-membered ring for the second pair ((**b**), marked yellow), and a four- and eight-membered ring for the third pair ((**c**), marked magenta)

puter systems, the number of rings in the RCSR database is not given as a subscript but in parentheses, i.e., 6(2).6(2).6(2).6(2).6(2).6(2). If an angle is part of two rings of different sizes, then for each pair of opposite angles, the smaller one is specified first; for example, the vertex symbol of the four-fold coordinating vertices of the zeolite net **Ita** is 4.6.4.6.4.8 (◻ Fig. 9.17).

The angles of five-fold and higher coordinating vertices can no longer be grouped according to a reasonable scheme, so there is no fixed order according to which the angles and ring sizes should be examined and specified. They are first examined in any order and then sorted according to increasing ring size and increasing subscript indices. For example, the vertex symbol of the six-fold coordinating vertices (i.e., there are already 15 different angles [= pairs of edges] meeting at a vertex) of the uninodal network **hxg** is $4.4.4.4.4$ $.4.6_4.6_4.6_4.6_6.6_6.6_6.6_6.6_6.6_6$.

The third reason why the determination of the vertex symbol can be complicated is that sometimes there are no rings for certain angles of a given vertex. This is the case, for example, with the relatively simple, uninodal 6-c net **pcu**: There are twelve four rings (four each for the three perpendicular planes, ◻ Fig. 9.18), but for three pairs of edges, namely, those having a 180° angles, you cannot find any ring. Of course, you can find cycles – closed paths of edges of any length that have a common start and end point – but none of them, not even the shortest possible one, fulfill the definition of a ring, sometimes also called fundamental circuits. This can also be expressed that along the cyclic path, there must be no shortcut (of exactly one edge) to the start vertex to be a ring. This is illustrated for one of the angles of the **pcu** net, formed by the edges a and b, in ◻ Fig. 9.18d. For the angles for which no ring can be found, we write as symbol an asterisk "∗." The complete vertex symbol of the uninodal 6-c net **pcu** is therefore 4.4.4.4.4.4.4.4.4.4.4.4.∗.∗.∗.

In nets with 6-coordinating vertices, we have to specify already 15 angles. It is easy to see that the number increases very rapidly with increasing coordination numbers and the corresponding notation becomes very unwieldy. For vertices with a coordination number of seven or larger, you should refrain from specifying the vertex symbol.

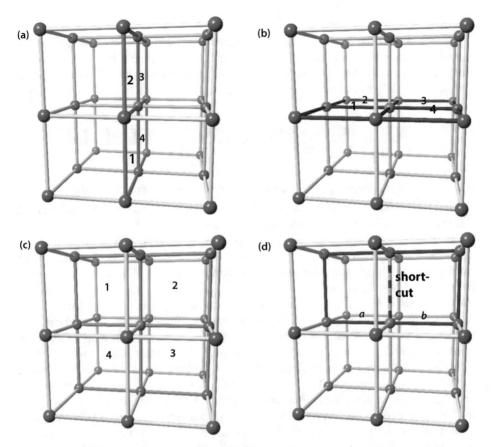

Fig. 9.18 Section of the **pcu** network. (a–c) The twelve four-membered rings originating from the vertex highlighted in orange. No vertex symbol can be specified for the three edge pairs that form an angle of 180°, since there is no ring for these edge pairs. This is illustrated for one of the 180° angles formed by the edges *a* and *b* in (d): The purple cycle contains a potential shortcut to the home vertex. Such cycles do not meet the definition of a ring. (Reprint courtesy of Wiley-VCH, from [3])

If all kinds of vertices of a net are specified according to the vertex symbol, is the net uniquely characterized? This is almost (but unfortunately only almost) the case. For example, there are the networks **dia** and **lon**, both uninodal networks with four-fold coordinating vertices, whose vertices in both cases have the symbol $6_2.6_2.6_2.6_2.6_2.6_2$. Interestingly, also the coordination sequence alone is not a distinctive feature of a net. If we consider both characteristics together, i.e., both the coordination sequence and the vertex symbols, then there is almost an unambiguity: With the two nets **edq** and **cdj**, there is only one exception. Both nets have identical vertex symbols *and* coordination sequences – however, they do have different symmetry.

vertex	CS_1	CS_2	CS_3	CS_4	CS_5	CS_6	CS_7	CS_8	CS_9	CS_{10}	cum_{10}	vertex symbol
V1	4	12	24	42	64	92	124	162	204	252	981	6(2).6(2).6(2).6(2).6(2).6(2)

9.3.1.8 Specification of the Edges

In the fifth table of the database entry for a net, the crystallographic specifications of the edges are given, again with their fractional coordinates (which refer to the center of the edge), the Wyckoff letters, the multiplicities, and the site symmetry. The coordination number of the edges is of course not given, because it is always two; an edge connects exactly two vertices.

edge	x	y	z	symbolic	Wyckoff	symmetry
E1	0.0000	0.0000	0.0000	0, 0, 0	16c	-3m

9.3.1.9 Tilings, Face Symbols, and the Concept of Transitivity

The last table refers to the tiling of the net and its so-called transitivity. What is behind these two concepts? First, a few introductory remarks.

The most obvious difference between classical inorganic crystal structures on the one hand and zeolites, MOFs, and other framework compounds on the other hand is their density. Zeolites, but especially MOFs, are typically materials with a low or even extremely low density, at least once the solvent is removed still remaining in the cavities after synthesis. If we look at MOFs as networks consisting only of geometric or topological points and lines having no extension whatsoever, there is not much of them left – and what we have in front of us can be considered as fragments of empty space. An interesting question in this context is: How can we fill this fractured space with geometric bodies in a way that the boundaries of the net are boundaries of this geometric body? The next question would be: To what extent could this space be filled, and how will the body look like to achieve the highest amount of space-filling? These issues of space-filling have a long tradition, beginning with the conjecture by Kepler from 1611 that there is no denser packing of (an unlimited number of) equal spheres than the cubic closest packing, which was proven only very recently [7]. Fedorov in turn could show in 1885 that there are only five types of convex polyhedra with opposite faces being parallel (so-called parallelohedra) that completely fill the space by translation alone [8]. The answers to the preceding questions concerning nets were given by the remarkable work of Delgado-Friedrichs et al. in 1999 by introducing the concept of tilings of space that allow to derive nets [9]. It is also said that the tiling "carries" the net. This tiling makes a periodic division of space, using a special type of face-sharing tiles that are generalized polyhedra. They are special because, for example, they do not have to be necessarily convex and the surfaces of the polyhedron do not necessarily have to be flat but may also be curved. But they are – juxtaposed – completely space-filling. The tiling for the **dia** net is illustrated in ◻ Fig. 9.19.

Basically, for a given net, there is more than one possible tiling that completely fills the space, depending on which rings you select that define the boundary faces of the 3D tiles. Therefore, rules for the tiling have been developed to ensure that the tiling for a given net is unique and unambiguous. Such tiling is called *natural tiling*. The property of a tiling to be unambiguous or unique allows to derive the net in an unambiguous manner: The edges of the net are given by the lines where at least three faces meet and the vertices by the points where at least three edges meet.

What can we do with these somewhat idiosyncratic tiles, and what are they useful for? First, they are very practical, because they allow to identify the different types of pores/

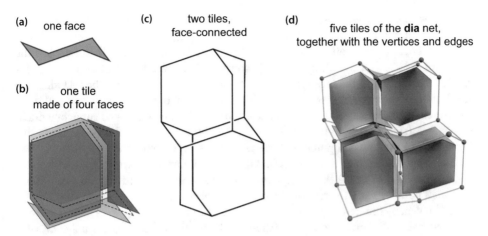

(a) one face

(b) one tile
made of four faces

(c) two tiles,
face-connected

(d) five tiles of the **dia** net,
together with the vertices and edges

◻ **Fig. 9.19** Illustration of the tiling of the **dia** net. Four non-planar faces (**a**) form a tile (**b**); several tiles are laid together face-by-face (**c**). (**d**) The tiling "carries" the net, which consists of the respective vertices and edges. (Reproduced with the kind permission of Wiley-VCH, from [3])

◻ **Fig. 9.20** The tiling of the zeolite LTA, which is composed of double-four rings (abbreviation, D4R, green, face symbol [4^6]), truncated octahedra (orange, face symbol [$4^6.6^8$]), and truncated cuboctahedra (blue, face symbol [$4^{12}.6^8.8^6$]). (Reproduced with kind permission of Verlag Wiley-VCH, from [3])

cages present in a porous crystal very quickly and to visualize them with the help of suitable software like GAVROG and ToposPro mentioned above. This is by no means limited to the pore types that can be described by classical convex polyhedra, such as zeolites. ◻ Figure 9.20 shows the natural tiling of the zeolite LTA; the creation of such an image takes only a few minutes. The generalized approach of natural tiling also makes it possible to visualize pores that are non-convex or irregular (in a not strictly mathematical sense), as is often the case with MOFs.

For the specification of the tiles, also a symbol is used. Because this symbol encodes the faces of which the tiles are made up, it is called *face symbol*. To avoid confusion with vertex symbols, the face symbol is placed in square brackets. The general notation of a face symbol is [$a^m.b^n.c^o...$], which means that the tile consists of m faces representing a polygon with a vertices, n pieces that are b-sided faces, and o faces that are c-gons, with $a < b < c$. An octahedron, for example, consists of eight triangles, so the face symbol is $[3^8]$. A cuboctahedron consists of eight triangles and six squares and therefore gets the face symbol $[3^8.4^6]$. If the tiling of a net is composed of several kind of tiles, each tile is specified according to the face symbol, and the ratio of occurrence can be included in a stoichiometric-like notation with a subscript: The tiling of the LTA zeolite (Fig. 9.20) is composed of double 4-rings (cubes), truncated octahedra, and truncated cuboctahedra in the ratio 3:1:1 giving the face symbol $[4^6]_3[4^6.6^8][4^{12}.6^8.8^6]$. In software packages like ToposPro, it is written computer readable as $3[4\wedge6] + [4\wedge6.6\wedge8] + [4\wedge12.6\wedge8.8\wedge6]$.

The second advantage of tilings is that they allow a further classification and systematic enumeration of nets – this is expressed by the so-called transitivity and comprises a four-digit number $pqrs$. The transitivity of a tiling specifies:

- Of how many different types of tiles the tiling consists of (s)
- How many different types of faces (these can be regular or non-regular polygons with flat surfaces or also curved fractions of a surface) occur in the tiling (r)
- How many distinct kinds of edges the faces have (q)
- How many different kinds of vertices the faces have (p)

The usage of "kinds" or "types" of tiles/faces/edges/vertices always refers to the number of edges/vertices, which are not symmetry-related to each other.

For the entry of the **dia** net in the RCSR, you will find in the last table that the transitivity is $pqrs = 1111$, i.e., there is only one type of tile, which is made up of identical faces, and these faces consist of symmetry-related edges and vertices. Nets with the transitivity 1111 are called *regular nets*. Transitivity can be regarded as a measure of the homogeneity or uniformity of a net. The larger the parameters $pqrs$, the more heterogeneous the net is.

9.4 The Principle of Minimal Transitivity

231 zeolite types are known, and about 3,000 nets have been described for MOFs. An interesting question regarding the transitivity is: How often do the different types occur? Are they equally distributed? Do certain types occur more often and others less frequently? That is indeed the case: Remarkably, the vast majority of MOFs synthesized so far form only a very limited number of different networks, which are also relatively simple, i.e., those with relatively small values for the transitivity. Based upon this observation, the principle of minimal transitivity was formulated. It is based on theoretical considerations related to the following question: If we have a certain number of *chemically different nodes* in our structure that can represent either an inorganic SBUs or branching point of the linker, how many different kinds of edges do we need to form a continuous network, i.e., to connect all nodes? Interestingly, it was observed that the large majority of MOFs that have been realized so far have underlying nets with a minimum number p of topologically different vertices and that for a certain value of p only a minimum number q of topologically different edges is required [8]. It should be noted that chemically different building units often result in topologically indistinguishable vertices, for example, in NaCl, which consists of two chemical building units but of only one type of octahedral 6-c node, so that NaCl forms topologically only a primitive cubic **pcu** network. Putting the principle of minimal transitivity in its simplest form, uninodal networks are much more frequent than

multinodal networks, and in binodal networks those with only one kind of edge are more frequent compared to those with two kinds of edges, and so on.

This concludes our little tour into the world of porous crystals and their topological description. Those of the readers who are further interested in these topics will find good starting points in References [3] and [9–12].

References

1. Online-Database of Zeolite Structures, IZA. http://www.iza-structure.org/databases/. Accessed 30 Sept 2019
2. Li H, Eddaoudi M, O'Keeffe M, Yaghi OM (1999) Design and synthesis of an exceptionally stable and highly porous metal-organic framework. Nature 402:276–279. https://doi.org/10.1038/46248
3. Hoffmann F, Fröba M (2016) Network topology. In: Kaskel S (ed) The chemistry of metal-organic frameworks. Wiley-VCH, Weinheim, pp 5–40. https://doi.org/10.1002/9783527693078.ch2
4. (a) Delgado-Friedrichs O, O'Keeffe M (2003) Identification of and symmetry computation for crystal nets. Acta Cryst A 59:351–360. doi: https://doi.org/10.1107/S0108767303012017; (b) http://gavrog.org/. Accessed 30 Sept 2019
5. (a) Blatov VA, Shevchenko AP, Proserpio DM (2014) Applied topological analysis of crystal structures with the program package ToposPro. Cryst Growth Des 14:3576–3586. doi: https://doi.org/10.1021/cg500498k; (b) http://topospro.com/. Accessed 14 of June 2019
6. (a) O'Keeffe M, Peskov MA, Ramsden SJ, Yaghi OM (2008) The Reticular Chemistry Structure Resource (RCSR) database of, and symbols for, crystal nets. Acc Chem Res 41:1782–1789. doi: https://doi.org/10.1021/ar800124u; (b) http://rcsr.net/. Accessed 30 Sept 2019
7. Delgado-Friedrichs O, Dress AWM, Huson DH, Klinowski J, MacKay AL (1999) Systematic enumeration of crystalline networks. Nature 400:644–647. https://doi.org/10.1038/23210
8. Li M, Li D, O'Keeffe M, Yaghi OM (2014) Topological analysis of metal–organic frameworks with polytopic linkers and/or multiple building units and the minimal transitivity principle. Chem Rev 114:1343–1370. https://doi.org/10.1021/cr400392k
9. Carlucci L, Ciani G, Proserpio D (2007) Networks, topologies, and entanglements. In: Braga D, Grepioni F (eds) Making crystals by design – methods, techniques and applications. Wiley-VCH, Weinheim, pp 58–85
10. Blatov VA, Proserpio DM (2011) Periodic-graph approaches in crystal structure prediction. In: Oganov AR (ed) Modern methods of crystal structure prediction. Wiley-VCH, Weinheim, pp 1–28
11. Batten SR, Neville SM, Turner DR (2009) Nets: a tool for description and design. In: Coordination polymers – design, analysis and application. The Royal Society of Chemistry, Cambridge, UK, pp 19–58
12. Batten SR (2010) Topology and interpenetration. In: MacGillivray LR (ed) Metal-organic frameworks: design and application. Wiley, Hoboken, pp 91–130

Supplementary Information

© Springer Nature Switzerland AG 2020
F. Hoffmann, *Introduction to Crystallography*, https://doi.org/10.1007/978-3-030-35110-6

Appendix

A.1. Solutions to Exercises

A.1.1. Solution to Exercise of ▶ Chap. 3

The solution of the exercise of ▶ Chap. 3 (◘ Fig. 3.18) is shown in ◘ Fig. A.1.

A.1.2. Solution to Exercise of ▶ Chap. 5

The solution of the exercise of ▶ Chap. 5 (◘ Fig. 5.7), where you were asked to draw locomotives at their new positions after applying a glide plane c, is shown in ◘ Fig. A.2.

A.1.3. Solution to Exercise of ▶ Chap. 6

In ▶ Chap. 6, ◘ Fig. 6.13, you were asked to identify the asymmetric unit or fundamental region of a chessboard. The small colored triangles in ◘ Fig. A.3 each mark asymmetric units. The plane group is $p4mm$. In the lower part of the chessboard, the unit cell is outlined in red. At the center there is a four-fold axis of rotation. The mirror planes shown as cyan lines must represent boundaries for the asymmetric unit.

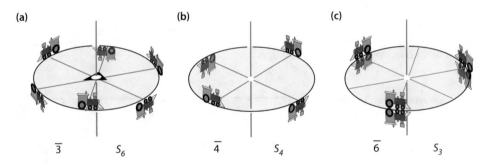

(a)   (b)   (c)

$\overline{3}$ S_6 $\overline{4}$ S_4 $\overline{6}$ S_3

◘ **Fig. A.1** Three arrangements of locomotives with different kinds of rotoinversion and rotary reflection axes. (**a**) Three-fold rotoinversion axis and six-fold rotary reflection axis, (**b**) four-fold rotoinversion as well as rotary reflection axis, (**c**) six-fold rotoinversion axis and three-fold rotary reflection axis

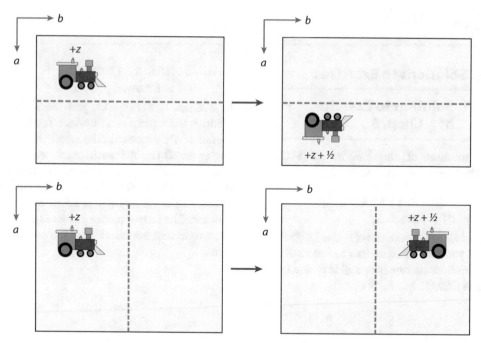

Fig. A.2 (left) The starting configurations of the locomotives for which a glide plane *c* is to be applied, with a mirror component that should be either the (*b, c*) (top) or (*a,c*) (bottom) plane. (right) The new locations of the locomotives after applying the glide plane *c*

Fig. A.3 Asymmetric units (colored triangles) and the unit cell (outlined in red, four-fold axis of rotation in the center, mirror planes are drawn in cyan) of a chessboard

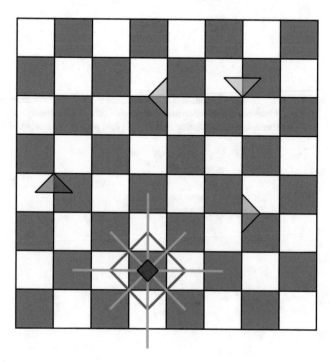

A.2. List of Minerals/Compounds (◻ Table A.1)

◻ **Table A.1** Minerals and compounds that are available as CIF or VESTA files at the accompanying website at ▶ https://crystalsymmetry. wordpress.com/textbook/

Name	Formula	Space Group
Azurite	$Cu_3(OH)_2(CO_3)_2$	$P2_1/c$
Barium titanate	$BaTiO_3$	$P4mm$
Barium titanate	$BaTiO_3$	$Pm\bar{3}m$
Baryte	$BaSO_4$	$Pbnm$
Benzene	C_6H_6	$Pbca$
Chalcanthite	$CuSO_4 \cdot 5\,H_2O$	$P\bar{1}$
Corundum	Al_2O_3	$R\bar{3}c$
Covellite	CuS	$P6_3/mmc$
CuBTC	$Cu_3(btc)_2$	$Fm\bar{3}m$
Diamond	C	$Fd\bar{3}m$
Dolomite	$CaMg(CO_3)_2$	$R\bar{3}$
Dumortierite	$Al_7O_3(BO_3)(SiO_4)_3$	$Pmcn$
Epidote	$Ca_2Al_2FeSi_3O_{12}OH$	$P2_1/m$
Ethylene	C_2H_4	$P2_1/n$
Fluorapatite	$Ca_5(PO_4)_3F$	$P6_3/m$
Fluorite	CaF_2	$Fm\bar{3}m$
Galena	PbS	$Fm\bar{3}m$
Gypsum	$CaSO_4 \cdot 2\,H_2O$	$C2/c$
Graphite	C	$P6_3/mmc$
Halite	$NaCl$	$Fm\bar{3}m$
Haüynite	$Na_3Ca(Si_3Al_3)O_{12}SO_4$	$P\bar{4}3n$
Hemimorphite	$Zn_4Si_2O_7(OH)_2 \cdot H_2O$	$Imm2$

◻ **Table A.1** (continued)

Name	Formula	Space Group
Ice Ih	H_2O	$P6_3cm$
Ilmenite	$FeTiO_3$	$R\bar{3}$
Magnesite	$MgCO_3$	$R\bar{3}c$
Magnesium	Mg	$P6_3/mmc$
Magnetite	Fe_3O_4	$Fd\bar{3}m$
Malachite	$Cu_2(OH)_2CO_3$	$P2_1/a$
MOF-5	$Zn_4O(bdc)_3$	$Fm\bar{3}m$
Muscovite	$KAl_2(AlSi_3O_{10})(F,OH)_2$	$C2/c$
Nickeline	$NiAs$	$P6_3/mmc$
Olivine	Mg_2SiO_4	$Pnma$
Pentagonite	$Ca(VO)(Si_4O_{10}) \cdot 2\,H_2O$	$Ccm2_1$
Perovskite	$CaTiO_3$	$Pbnm$
Pyrite	FeS_2	$Pa\bar{3}$
Quartz	SiO_2	$P3_221$
Siderite	$FeCO_3$	$R\bar{3}c$
Sphalerite	ZnS	$F\bar{4}3m$
Struvite	$NH_4MgPO_4 \cdot 6\,H_2O$	$Pmn2_1$
Talc	$Mg_3Si_4O_{10}(OH)_2$	$C\bar{1}$
Tausonite	$SrTiO_3$	$Pm\bar{3}m$
Tellurium	Te	$P3_121$
Topaz	$Al_2SiO_4F_2$	$Pbnm$
Turquoise	$CuAl_6(PO_4)_4(OH)_8 \cdot 4\,H_2O$	$P\bar{1}$
Wollastonite	$CaSiO_3$	$P\bar{1}$
Vivianite	$Fe_3(PO_4)_2 \cdot 8\,H_2O$	$C2/m$
Zeolite LTA	SiO_2	$Pm\bar{3}m$
Zeolite SDA	SiO_2	$Im\bar{3}m$

A.3. Supplementary Materials (Online)

At the accompanying website to this book (▶ https://crystalsymmetry.wordpress.com/textbook/), the following additional materials can be found:

— 32 sheets for paper models of crystals according to the 32 crystal classes

— Poster of crystals classes (DIN A0 format).
— A video with Crystallographic poetry.
— Animations of the glide plane operations a, b, c and n, d, e.
— Animations of the 2_1 screw rotation and glide reflection n of ethylene.

Index

Printed in the United States
by Baker & Taylor Publisher Services